DYNAMIC MODELS FOR THE INTER-RELATIONS OF REAL AND FINANCIAL GROWTH

INTERNATIONAL STUDIES IN ECONOMIC MODELLING

Series Editor

Homa Motamen-Scobie

Executive Director
European Economics and Financial Centre
London

Economic Modelling in the OECD Countries
H. Motamen-Scobie

Modelling the Labour Market
M. Beenstock

Input-Output Analysis
M. Ciaschini

Models of Disequilibrium and Shortage in Centrally Planned Economies
C. Davis and W. Charemza

Economic Modelling at the Bank of England
S. G. B. Henry and K. D. Patterson

Recent Modelling Approaches in Applied Energy Economics
O. Bjerkholt, Ø. Olsen and J. Vislie

International Commodity Market Models
O. Güvenen, W. C. Labys and J. B. Lesourd

Dynamic Models for the Inter-relations of Real and Financial Growth
H. Ekstedt and L. Westberg

Dynamic Models for the Inter-relations of Real and Financial Growth

Hasse Ekstedt

Assistant Professor, School of Public Administration, Göteborg University, Sweden

and

Lars Westberg

Associate Professor, School of Economics, Göteborg University, Sweden

CHAPMAN & HALL

London · New York · Tokyo · Melbourne · Madras

UK	Chapman & Hall, 2–6 Boundary Row, London SE1 8HN
USA	Chapman & Hall, 29 West 35th Street, New York NY10001
JAPAN	Chapman & Hall Japan, Thomson Publishing Japan, Hirakawacho Nemoto Building, 7F, 1-7-11 Hirakawa-cho, Chiyoda-ku, Tokyo 102
AUSTRALIA	Chapman & Hall Australia, Thomas Nelson Australia, 102 Dodds Street, South Melbourne, Victoria 3205
INDIA	Chapman & Hall India, R. Seshadri, 32 Second Main Road, CIT East, Madras 600 035

First edition 1991

Typeset in 10/12 Sabon by
KEYTEC, Bridport, Dorset
Printed in Great Britain by
T. J. Press (Padstow) Ltd, Padstow, Cornwall

ISBN 0 412 35350 4

British Library Cataloguing in Publication Data
Ekstedt, H.
Dynamic models for the inter-relations of real and financial growth.
I. Title II. Westberg, L.
339.5

ISBN 0-412-35350-4

Library of Congress Cataloging-in-Publication Data
available

Contents

Preface

The high growth rate in many industrial countries during the first two decades after the second world war produced a financial expansion, which in its turn made possible an extension of the growth process. But it also led to a financial crisis, when increasing costs during the 1970s created growing deficits in the balance of payments and the government budgets in many countries.

This development brings to the fore a disequilibrium approach for analysing the inter-relation between real and financial development, in a world where agents are not able to realize their plans. This inter-relation makes it natural for us to start in the theoretical world of Keynes and the Stockholm school; since the modern disequilibrium analysis often disregards the financial side of the economy.

In our approach we are influenced by the Cambridge school and the French disequilibrium economists, J. M. Grandmont, E. Malinvaud and J-P. Benassy to mention some. The works of H. Minsky, G. R. Gurley, J. G. Shaw and A. Enthoven have also been important for us.

The book is a joint effort and we are both reponsible for all errors. In actual writing one author had the initial responsibility for each chapter which was then subject to criticism. Ekstedt wrote the initial drafts of chapters 2, 4 and 8 and Westberg those of 3, 5 and 6 while both wrote sections of chapters 1 and 7.

We have incurred a very special debt to the late Profesor Tor Fernholm whose ideas of extensions of the input-output system to include financial flows and stocks have inspired us.

Professor Börje Kragh and Dr Karl-Olof Faxén have read preparatory works and supported us with stimulating criticism.

Senior lecturer Carl-Gustav Ivarsson at the School of Public Administration at the University of Göteborg has read parts of the manuscript and has made valuable comments.

We are particularly grateful to Dr Homa Motamen-Scobie, whose support made this work possible.

We are greatly indebted to the staffs of the School of Public Administration and the Department of Economics at the School of Economics and Commercial Law, the University of Göteborg, which have supported the preparation of this book.

We are glad to acknowledge the indispensable aid of Mrs Birgit Lundbladh, who contributed not only with her great ability as a typist, but also with her unlimited patience in interpreting our handwriting and incorporating innumerable revisions on the occasions when it was needed.

We are also much indebted to Miss Elisabeth Linsten who has drawn the figures with great skill.

1

Introduction

1.1 DISEQUILIBRIUM AND UNCERTAINTY

Some years ago a list of macroeconomic models developed in the OECD countries appeared. For each model it was reported whether it was of the equilibrium or disequilibrium type, the number of equations and the number of exogenous variables. In this list we found models which contained up to 25 000 equations, probably covering the most intricate part of the economy which they were modelling.

These models are in some sense general, as they account for almost all possible relationships in a modern economy. However, they are highly aggregated since agents are lumped together as average consumers, investors, wealth owners etc. They also include relations between market prices under the assumption of some specific characteristics of the agents involved.

As a contrast to these empirical models we can analyse the existence of equilibrium in an abstract market economy consisting of agents who are represented by their preference relations, which are assumed to be continuous, monotone and strictly convex, and their initial endowments, which are defined in the positive real goods space (Debreu, 1974). An analysis of this kind can also be called general, but in a different sense to the analysis described in the previous paragraph.

In the first example of economic analysis, the claim of generality comes from the fact that the model builder has considered most of the important economic variables and defined them in equations which together define the whole economy. However, the equations are normally estimated using historic material and are therefore some kind of average relation for their respective part of the economy during the relevant period. This means that an estimated model which expresses some average for a time period of an empirical economy does not represent the same economy today or in the future. There exists no model which is correct on the average. A random external event will always hit the model in some specific part. If this part is mis-specified relative to the economy in question or to other parts of the economy, the total result of the model analysis will be different to a greater or lesser degree from the reality. The ability of these models to forecast developments or analyse the effects of policy measures is normally good when the economy in question is stable and poor when the economy experiences some kind of structural change. As an econometrician said: 'When forecasts are possible to do they are not needed, but when they are needed they are not possible to do'.

The other approach is general in the sense that if the definition of agents is empirically correct and the axioms are fulfilled, the analysis will cover the totality of empirical possibilities. The weakness of this analysis in relation to empirical studies is of course that it demands very precise empirical knowledge of the preference structure and the environment of the market. Furthermore, the existence of equilibrium does not imply anything about the process of convergence from one equilibrium to another if several equilibria exist, or from a disequilibrium situation to an equilibrium. If the convergence time is not negligible we are supposed to construct a theory of economic behaviour in states of disequilibrium. However, it is obvious that few of the empirical models of this type are attempts to operationalize the theoretical model of general equilibrium.

Our primary interest in this book is the dynamic behaviour of an economy in disequilibrium. We shall employ an approach somewhere between the two discussed above. On the one hand we utilize econometric studies of the Swedish economy based on a dynamic macroeconomic model. The essence of this model is to describe general restrictions placed on various aggregates of agents. On the other hand we employ the assumed generic behaviour of the agents in general equilibrium theory. The application of this behaviour to disequilibrium situations will lead to successive changes in the aggregate restriction structure, which will appear as parameter changes in the model.

During the last two decades there has been much argument about the microeconomic foundations of macroeconomic theory. It is important to realize that this does not mean that macroeconomic theory should converge to general equilibrium theory. The only time when this is relevant is when we actually face a general equilibrium. The reason why we normally have a difference between the microeconomic and macroeconomic levels of the economy is that the economy is in a state of disequilibrium. The aggregate behaviour of the economy will cause failures to some agents relative to their *ex ante* expectations. In an equilibrium situation the initial endowments of the agents can be taken as given and then the behaviour of the agent can be focused on the marginal decision. In disequilibrium, however, the initial endowments of the agents will undergo revaluations and thereby create non-marginal changes in the behaviour. The main problem of macroeconomic theory since Keynes has been to analyse behaviour relative to changing valuations of capital and labour under short-run liquidity restrictions.

However, it is important to determine whether the marginal behaviour of different situations will be of a stabilizing or a destabilizing character. In disequilibrium, where the agents trade at false prices, in general several possible equilibria are attainable, given a time sufficient for the economy to converge. In this situation the individual agent has to choose between contradictory or even contrary strategies. An example is when there is a severe fall in equity prices. Depending on the liquidity and the expectations of future price development, the agent may choose either to sell the equities and face a windfall loss or to keep the equities.

Another feature of the discussion of the microeconomic foundations of macroeconomic theory has been the varying attitude towards institutional restrictions of different kinds. Sometimes one receives the impression that disequilibria are created by restrictions applied by public authorities. However, we believe that disequilibrium is the normal situation in empirical economies and that it is created by deficient flexibility of factors of production combined with current liquidity restrictions and uncertainty. Whether the public restrictions worsen or ease the disequilibrium is a matter for empirical enquiries. We feel that economists sometimes treat general equilibrium analysis more as a platonic ideal world that we should strive for than an intellectual tool.[1] One often meets the attitude that institutional restrictions are, at best, a necessary evil which makes it impossible for the economy to reach some first-best equilibrium. As we shall see, this attitude provides a non-cultural and non-historical way of analysing economic behaviour. Economics then becomes a collection of divine eternal rules which are separate from human civilization. We do not accept this platonic view but insist on the possibility of collective actions where the individual preference relations also include preferences about the functioning of the economic system (Sen, 1976).

It is theoretically possible to attach a general equilibrium to a given arbitrary specification of an empirical economy. The problem is that the process of convergence to a general equilibrium implies that the market process must be neutral with respect to the agents involved and their respective initial endowments. In general equilibrium theory this is guaranteed by assumptions about competitive markets and the information structure.

In an empirical economy there are many degress of competitiveness of different markets and there are more restrictions on the agents than just the initial endowments which primarily consist of assets of different kinds and labour with different technical specifications. These restrictions can be of many kinds. The flexibility of capital assets and labour may vary, and liquidity constraints may vary among the individuals and over time.[2] There

[1] Two interesting articles on this subject are In Praise of Economic Theory in Hahn (1985) and Amoroso (1937).

[2] Observe that a liquidity restriction implies that some goods are necessary in the present time irrespective of the long-term effects.

are also social and other types of institutional restrictions which are binding. The most important restriction in an empirical economy is probably the uncertainty of the future combined with current liquidity restrictions.

In an economy where there are a large number of varying kinds of restrictions and uncertainties about the future, many possible convergence processes to a number of equilibria could be going on at the same time. Actually, as Keynes points out, productive capital of long duration is the bearer of a specific equilibrium which is attached to the state of expectations at that particular time when it was installed. Every revaluation of this capital stock will then theoretically point towards some other general equilibrium which is different from the *ex ante* equilibrium. The reason why the situation does not collapse into chaos is in fact the large number of restrictions which keep the behaviour of the agent within rather narrow borders. We could therefore speak of a stable situation without general equilibrium. Of course we could assume stable periods when we have stable growth and stable price relations. These periods may exhibit many of the same characteristics as general equilibrium growth. However, as long as it is not proved that natural resources are eternal and the waste from human activities is negligible we shall always have the possibility of future revaluations which sooner or later will have effects on current prices. In this case the assumed steady state equilibrium growth will show up as a false description.

We are also interested in investigating why some economies are stable and some are unstable. We can speak of relative stability when capital assets of different kinds and the labour of different agents do not undergo substantial revaluations. Note that we do not speak about relative prices of final goods. When we have changes in the relative prices of final goods but no revaluations of the 'initial endowments' it means that the productive resources are sufficiently flexible not to change the relative wealth in its broadest sense.

As mentioned above we are interested in the dynamic behaviour of disequilibrium situations. Let us exemplify this more closely with the help of the classical 'Stockholm school' *ex ante–ex post* analysis (Lindahl, 1939). Assume a rational agent, rational in the ordinary sense of economic theory. He acts *ex ante* with respect to his understanding of the restriction space and tries to maximize some outcome of his actions. At the same time, however, his fellow agents are doing the same with their respective conceptions of the restriction space. They will also interpret our agent's actions with respect to their specific conceptions of the restriction space which could be different from the conception of our agent. As long as we are not discussing perfect competition with uniform information, this will usually mean that our agent's actions will turn into a different result *ex post* from that which he had planned *ex ante*. This situation will now lead on the one hand to revaluations of durable stocks of capital and on the other to revisions of actions. The revaluation of stocks will lead to, at least, revisions of the perceived restrictions, particularly concerning liquidity, in relation to a new *ex ante* decision.

The situation indicated in the above example would not be a problem if resources were perfectly flexible and the time schedules of the restrictions on the agents involved were identical. The problem in an empirical economy is that the long-term decisions are usually displayed in the creation of inflexible assets which place severe restrictions on future actions. Indeed, the most important restrictions are always short-term liquidity restrictions.

When we speak of liquidity restrictions we usually think of the demand for the medium of exchange. However, there will be restrictions of the same kind in a barter economy. This will mean that the agent partly has a lexicographic preference order, i.e. the individual demands a certain quantity of some good or goods irrespective of the relative price or, in our case, of future consequences (Fig. 1.1). We could simplify the argument by saying that in order to survive in the long run it is certainly necessary to survive in the short run.

A long-term decision which diminishes the flexibility implies that the future short-term restrictions can be different from those expected when the decision was made. In general equilibrium theory we normally escape this problem by assuming competitive markets and a perfect or uniform information structure. Of course, in theory we could always specify the restriction system within the initial endowment vector. This is normally implied by the assumption of perfect information and perfect future markets. Radner (1968, 1974, 1982) enlarged the definition of the consumer in general equilibrium analysis when he introduced a vector of information in addition to the vector of initial endowments. The equilibrium is then dependent not

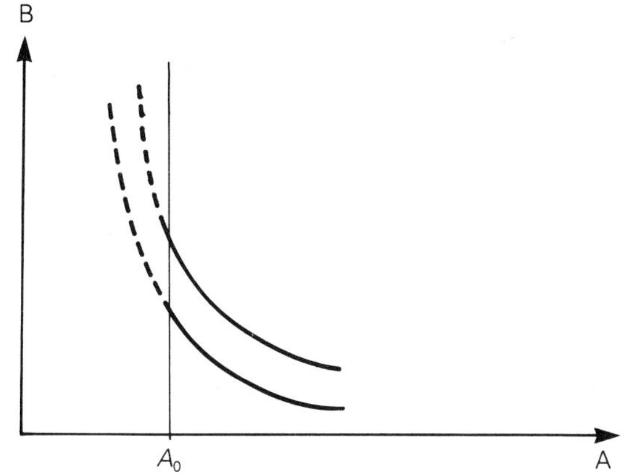

Fig. 1.1 Lexicographic preferences: the individual is forced to have the quantity A_0 of good A irrespective of the price relation.

only on the distribution of endowments but also on the distribution of information. However, when we make the theory operational we must not assume away the operational difficulties and still maintain the empirical relevance of the analysis. This should be one of the elementary results of the theory of second best.

The existence of short-term binding liquidity restrictions and the inflexibility of resources would not be a problem if there were no uncertainty about the future. The word 'uncertainty' is sometimes confused in economic literature. We shall follow the views of Knight (1921) and Keynes (1922).

An economy as a whole can be considered in four different ways. First, we have the situation where all the agents involved have identical subjective distributions over the objective space of possible outcomes. Second, all agents conceive the same space of possible outcomes, which is identical with the objective space, but they have different subjective probability distributions. If the second case exists at the same time as we have a competitive market and flexible resources, for example, on the security market, there will be some subjective probability distribution that is better than the others. It is then reasonable to expect, given rational agents, that the subjective probability distributions will converge through the information implicit in the price formation. Third, the agents have different conceptions of the possible space of outcomes, and therefore different probability distributions, but the union of all individually perceived spaces of outcomes includes the totality of possible outcomes. If we assume rational agents and competitive markets it should be possible for the agents to also arrive at the same probability distribution in this case, given a stable environment. The fourth situation, which we believe to be the normal empirical *ex ante* situation, occurs when the individually perceived spaces of outcomes are different from each other, as in the third case, but the union of all individual spaces does not include the totality of objectively possible outcomes.

We can translate these situations for the economy as a whole to individual situations.

1. The individual considers himself to be sure of the possible outcomes and has a subjective probability distribution over these outcomes.
2. The individual believes that he knows the possible space of outcomes but he is not able to assume any better probability distribution than the rectangular. This situation is worse than the first one because it means that there is less information about the outcomes. However, it is still better than the following situation.
3. The individual realizes that he does not know all the possible outcomes. In the last case in particular the expression 'degree of confidence' is two dimensional. On the one hand the agent has some kind of subjective probability ranking among the known possible outcomes, whereas on the

other he has to judge whether the acquired information is sufficient relative to other agents.

When we now compare the individual situations with the four types of situations for the economy as a whole it is obvious that it is possible for an individual to be in case 1 for the individual but in case 4 for the economy as a whole. It is also obvious that locally in time and space we could think of situations which approach type 1 for the economy as a whole and type 1 for the individual. A decision on the financial market for short-run securities in the case of stable real circumstances and well-informed agents is an example of this situation. It is also possible that price reactions promote the convergence of probability distributions.

Another example of a situation where the generality of the environment influences the degree of uncertainty can be found in the analysis by Bower (1970) and Nilsson and Persson (1988) of the relation between the actual outcome of an investment and the outcome expected at the time of the investment decision. Both groups examined three types of investments in production capital: cost-reducing investments, capacity-increasing investments and investments in new products. Their results are summarized in Table 1.1.

Two important questions with respect to markets arise when we introduce uncertainty of the future in the system. The first is whether the effects of the market process are neutral with respect to the agents when the initial endowments are different. Malkiel (1985) gives the following illustration of the risk situation (Fig. 1.2). The point is that risk premium is only relevant when the agents face unsystematic risks. That is, we assume that each agent has a rational diversification of his portfolio which is enough to eliminate the unsystematic risks. However, if we apply this reasoning to the economy as a whole, it is obvious that the possibility of diversification is a function of the distribution of initial endowments, i.e. income and wealth.[3]

Another question is whether the information inherent in the market price formation is efficient when the economy is in a situation of genuine uncertainty. The effect of discounting is that an asset yielding an expected

Table 1.1 Source: Nilsson and Persson, 1988

	Ratio of actual to expected outcome	
Type of investment	*Bower*	*Swedish firm*
Cost reducing	1.1	1.0
Capacity increasing	0.6	0.6
New product	0.1	0.5

[3] A simple example of this is if, when playing roulette, we play on dozens and double the stake each time we lose. The probability of losing the initial sum before winning the first time will be a decreasing function of the sum.

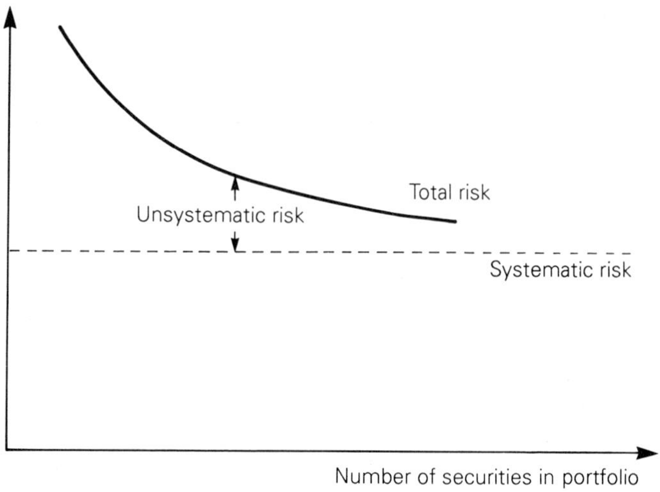

Fig. 1.2 Portfolio risk (standard deviation of return).

amount over several future periods will be transformed into a one-dimensional price and a shift in this price will have immediate consequences for wealth and liquidity. A discrepancy between *ex ante* and *ex post* proceeds will have two effects. On the one hand the agent must revise his expectations of the future proceeds; on the other he must revise his subjective probability distribution of future outcomes. This means that there are successive revisions of the systematic risk and therefore of the risk premium. The expression 'systematic risk' can be taken as the subjective probability distribution of exogenous risks relative to a specific state of information. The word 'uncertainty' refers to the agent's confidence in current information.

An increase in the uncertainty will on the one hand increase the claims of productive capital and thereby decrease the investment demand, and on the other increase the demand for liquidity. The increase in the claims of yield on productive capital will also induce revaluation of the existing stock of capital which, together with the increased demand for liquidity, will make floating fund finance more risky irrespective of the current flow of liquidity from production. The fundamental financial problem of the firm is that its capital assets are always valued at the market relative to the current state of expectation while its liabilities are normally financial contracts in nominal terms. An increasing stock of inflexible production capital may induce agents and public authorities, especially when there is little competition, to act in such a way that revaluations are avoided but the economy as a whole is driven into a more severe disequilibrium. An example is when the monetary authorities drastically increase liquidity to avoid breakdowns of the stock exchange and thereby increase the discrepancy between the market valuations of equities and the long-run yield of production capital.

From our earlier considerations we find that decisions made by individual agents which seem natural in the short run may alter the conditions of the economy in the long run. In a small open economy which has experienced rapid growth where the industry has increased its foreign sales the conditions for economic policy will continuously change. In Table 1.2 we compare two situations, assuming equivalent competitive conditions in the world market. It is obvious that a fiscal policy based on preserving high employment and demand will keep up the wage level but will also have positive effects on sales and profits in situation 1. However, from the point of view of the firm this is not self-evident in situation 2. The total risk of the company is clearly much higher in situation 2 and the effects of fiscal policy on the firm will be less satisfactory.

A central theme of our study is the reciprocal dependence between production and income formation on one side and financial development on the other. In the flowchart shown in Fig. 1.3 these interrelations are represented by arrows which link the price vector with the block for aggregate demand. In this block the investment is connected with the capital stock which, together with the labour force and technical changes, creates the full capacity product Y^*. The ratio of current production—a result of aggregate demand—to the full capacity product gives the capacity utilization which, in turn, influences productivity and factor income distribution as we have seen earlier.

As we proceed with the book we should bear in mind a dynamic model with the flowchart shown in Fig. 1.3. The flowchart can be divided into three blocks. The demand block consists of different parts of aggregate demand which create the total demand for goods and services. The production block contains the current stock of capital and the short-run supply of labour which we assume to be inelastic in the short run. The ratio of the actual demand to the potential supply, which is the level of production when the capital stock is fully utilized, gives the capacity utilization. The profit of the firms is treated as a residual; it is only in equilibrium that profit follows from the marginal conditions.

The variation in profits has two effects, which we shall analyse more closely later in this chapter: one is the revaluation of capital and the other is

Table 1.2

	Domestic market	*International market*
Situation 1		
Share of production (%)	75	25
Share of market (%)	35	$<$ 1
Situation 2		
Share of production (%)	17	83
Share of market (%)	25	$<$ 1

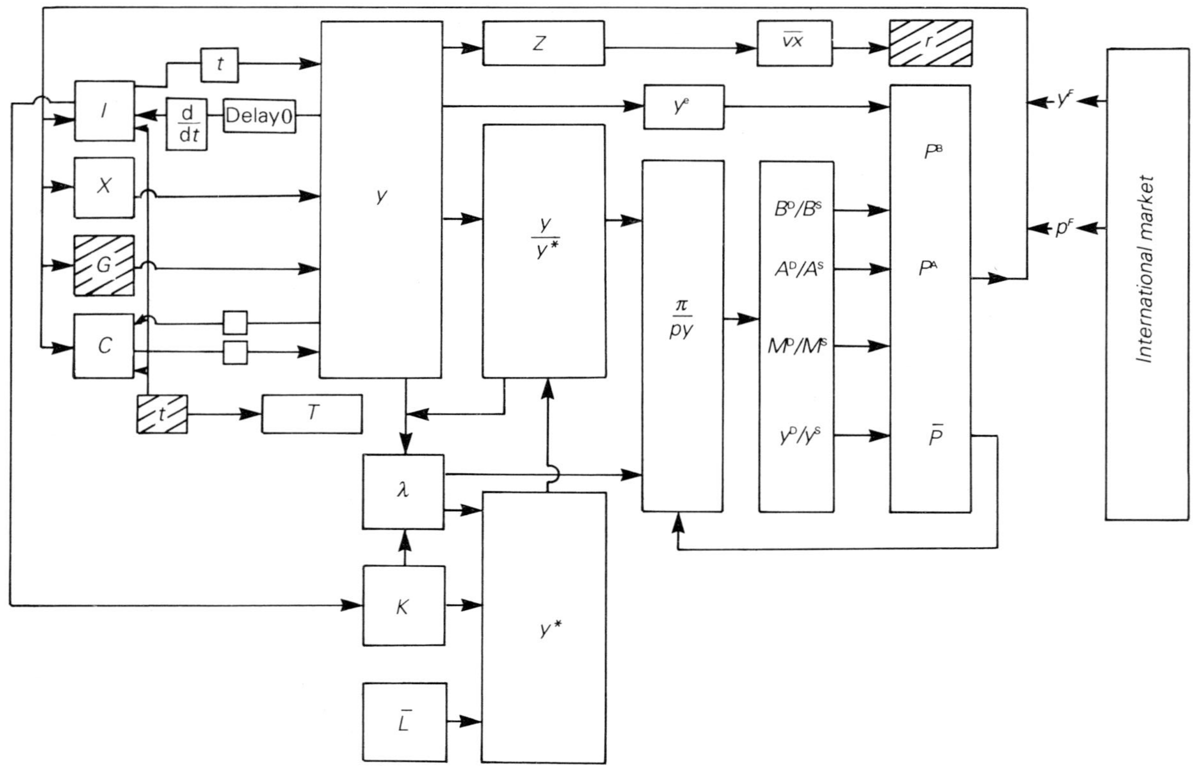

Fig. 1.3 Flowchart of an economy.

the change in the profit share. If we assume that capitalists and wage earners have different financial preferences, a change in the relative share of profits will alter the relative distribution of preferences on the financial market.

The third block is the price formation block. The acquired *ex post* situation will cause new *ex ante* behaviour and will thereby alter demand and supply in the real and financial markets. However, price formation will be affected by the development of foreign real and financial markets.

The discount rate r is assumed to be an instrument for monetary policy and therefore is influenced by the exchange rate $\overline{VX}$. In turn, the discount rate affects the short bond rate, and hence there is an arrow to the price vector, where bond prices P^{B} are located.

An important idea which the chart can illustrate is the central role of factor income distribution. Factor income distribution is the result of capacity utilization and growth in productivity, which is characterized by both cyclical and long-run variations. Changes in factor income distribution in turn influence supply and demand on the asset market. This flowchart can be used in conjunction with input–output technique which is discussed in the next section.

1.2 STOCKS AND FLOWS IN THE REAL AND FINANCIAL PARTS OF THE ECONOMY

1.2.1 Input–output systems

A central problem in our study is the interdependence between real and financial development in the economy. This interdependence leads both to short-term cyclical changes in production, prices and income and to changes in the medium-term and long-term growth rate in resources and their use for different types of finished products.

In order to study the total production system of the economy we introduce an input–output system with three matrices B, A and Y for finished goods, intermediary goods and primary inputs respectively. The primary inputs include domestic factor services and imports. The sum of finished and intermediary products—the gross product—is presented in the column vector X. The expenditures of the domestic production sectors for input goods are recorded in the columns of the B matrix for intermediary goods which is based on a functional sectorization of the production activity. The number of sectors can be quite large if the aim is to give a detailed account of the structural development of production. As we are interested in identifying only those types of structural change in production which may be important for factor income distribution and credit market conditions we can be satisfied with a more aggregated production sector

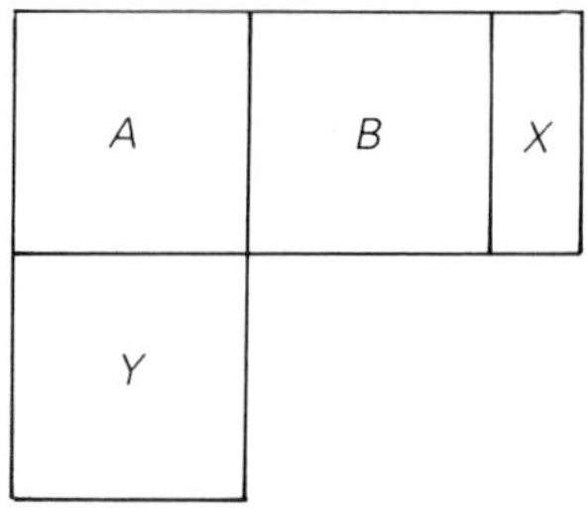

Fig. 1.4 Outline of an input–output system.

division. The basis for our division is, for instance, differences in productivity growth and in monopolistic situations, and different rules of the financing of expenditures in the private and government sectors.

In our analysis of the input–output system we are also restricted by institutional conditions, which are not present in the Walrasian world generally assumed in input–output analysis (Leontief, 1946). Therefore we assume that prices for goods and services are generally sluggish, in part depending on the existence of a monopolistic situation in most production branches. Of course, an exception has to be made for certain types of raw materials. We also assume that wage rates are generally contracted for a longer period than the extent of business production planning. Finally, it is of course a central assumption of input–output analysis that the technical conditions are given in so far as the need for intermediary goods is firmly related to the gross product of the production sector in question. The same can generally be assumed for the sector's need for labour and imports. The result of these technical and institutional restrictions is that, as prices and input costs are mostly given, it is the quantity of goods and services supplied during the period which determines the operating gross surplus of the production sector in question.

The role of input–output analysis is illustrated in Fig. 1.5 which shows the transactions table for an economy. The input–output system in the transactions table includes the four matrices Y, A, B and X discussed earlier. To the system can be added X', the transpose of the gross output X, which signifies the gross input and which according to the input–output analysis for each sector must be equal to the gross output. The A matrix is a quadratic matrix. The number of rows and columns corresponds to the number of domestic production sectors that is of interest for the analysis. Each row indicates the delivery of intermediary goods of a domestic production sector. The delivery of finished products from the same domestic production sectors that are included in the A matrix is accounted for in the rows of the B matrix. The sum of deliveries of intermediary goods and finished goods from the domestic production sectors is registered in the X vector for the gross product of the domestic production vectors.

The input–output analysis can inform us how the value of the output of

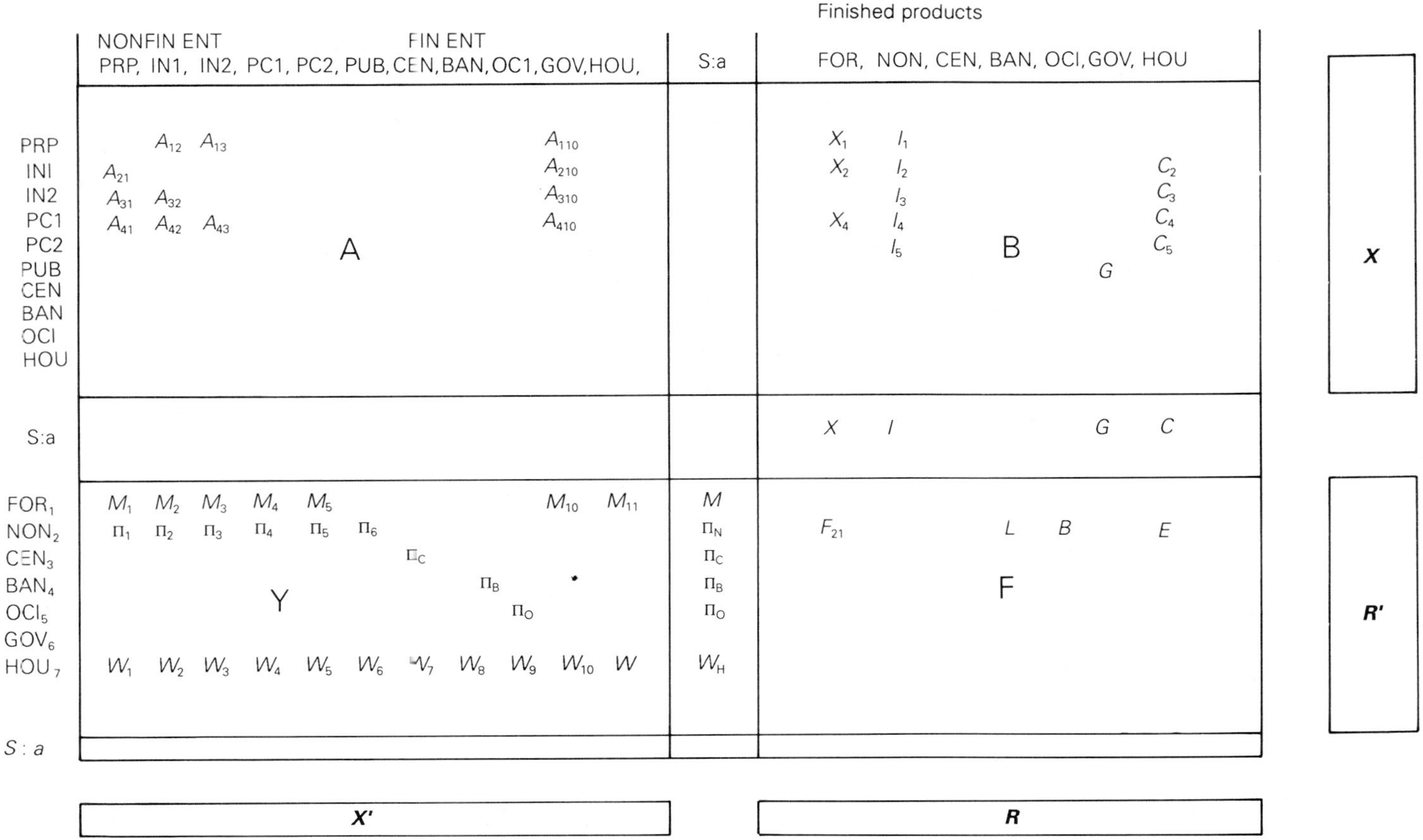

Fig. 1.5 Transactions table for flows: PRP, primary products sector (compare Kaldor); IN1, industry in K sector; IN2, industry in S sector; PC1, private service, industry-oriented; PC2, private service, household-oriented; PUB, public services; FOR, foreign sector; NON, non-financial sector; CEN, central bank; BAN, other banks; OCI, other credit institutes; GOV, government; HOU, households.

a sector, i.e. its gross output, is the result of different kinds of inputs. The sum of a sector's output registered in the X matrix is equal to the sum of the sector's inputs registered in the X' matrix. The inputs necessary for the production of a sector's gross product are therefore given by the columns of the sector included in the A and Y matrices. The cost for deliveries of goods and services from other domestic production sectors is registered in the columns of the A matrix, and the costs for imports of goods and services and for wages and salaries are registered in the columns of the Y matrix. If the gross output of the domestic production sector in question is higher than the input costs, the sector has received an operating gross surplus, which is accounted for in the sector's column and row in the Y matrix as a part of the period gross factor income.

In this way input–output analysis can be used to define a production sector's operating gross surplus. For sector i this definition can be written as follows:

$$\pi_i = P_i - a_{\mathrm{n}i}P_i - a_{\mathrm{m}i}P_{\mathrm{m}} - a_{1i}P_1 - a_{2i}P_2 - \ldots - a_{li}P_l$$

where

π_i is the operating gross surplus in sector i

P_i is the price of the gross product in sector i

$a_{\mathrm{n}i} = N_i/X_i$ is the ratio of the input of working hours in sector i to the gross product in sector i

$a_{\mathrm{m}i} = Z_i/X_i$ is the ratio of the input of imports in sector i to the gross product in sector i

$a_{1i} = A_{1i}/X_i$ is the ratio of the input of intermediary goods from sector 1 to sector i to the gross product in sector i

$a_{2i} = A_{2i}/X_i$ is the ratio of the input of intermediary goods from sector 2 to sector i to the gross product in sector i

$a_{li} = A_{li}/X_i$ is the ratio of the input of intermediary goods from sector L to sector i to the gross product in sector i

As it is assumed that prices are given, the ratio of a sector's operating surplus is given by the input coefficients for wage costs, import costs and costs for intermediary products.

Of course, this equality is the result of our definitions of the different outputs and inputs and does not give any information about the equilibrium in the sector. Neither does the equality between total inputs and total outputs in each sector. The only equilibrium concept which could be used here would be an equilibrium *ex post*, the notion of the Stockholm School which only involves the accountable equality between the transactions on a market after the end of the transaction period.

In order to analyse the *ex ante* equilibrium—if the vocabulary of the Stockholm School can be used—it is necessary to have information about the plans for the period under investigation in the various sectors of the economy. In order to be fungible the input–output system with its connections with Walrasian theory has to be supplemented with a macroeconomic theory, which works partly with a different type of sector from the production sector system used in the input–output system. The sector system connected to the macroeconomic model is based on an institutional division of the economy, where the demand for finished goods is divided between domestic and foreign demand, between demand for capital building and consumption, and between demand in the private and public sector. A new input–output system of a financial nature which completes the traditional real input–output system is constructed on the basis of this division of institutional sectors.

The financial input–output system is included in Fig. 1.5 in the four matrices B, F, Y and R. The financial output consists of the real and financial capital growth in the institutional sectors. It is registered (including depreciation) in the sector columns in the B and F matrices—real capital growth in the B matrix and financial capital growth in the F matrix. As well as capital growth, consumption expenditure is registered in the sector columns of the B matrix and transfer expenditure is registered in the sector columns of the F matrix. The foreign sector output registered in a column in the B matrix consists of exports of goods and services, which results in capital growth in foreign assets. The sum of the output of real and financial capital growth together with consumption and transfer expenditures is registered in row vector R.

The input to the financial input–output system consists of the factor income received by the institutional sectors and registered in their rows in the Y matrix. Therefore, while the columns of the Y matrix depend on the number of functional sectors in the production system, the number of rows is given by the number of institutional sectors.

The institutional sectors also receive financial input in the form of credits and transfer incomes. These inputs are registered in rows in the F matrix. The sum of the sectors' financial inputs in the form of factor incomes, transfer incomes and credits is registered in the column vector R'. The total financial input registered in this vector is of course equal to the total financial output registered in the row vector R.

Fig. 1.5 shows that the real and financial input–output systems overlap to some extent. This is the case for the B matrix and the Y matrix. The rows of the B matrix register the delivery of finished products which is part of the real input–output system. The columns of the B matrix indicate the expenditure on finished products in the institutional sectors which belong to the financial input–output system.

In the same way the Y matrix belongs to both the real and the financial input–output system. Its columns register the cost in the functional sectors for primary inputs plus the vectors' operating gross surplus. These transactions are part of the real input–output system. The rows in the Y matrix indicate the factor incomes earned in the institutional sectors. These incomes are part of the financial input–output system.

Finally, there are two matrices which are parts of only the real and the financial input–output systems respectively. The A matrix, in which the columns register expenditure of the functional sectors on intermediary products, indicates the technical relations between a given gross product and such expenditure on intermediary goods. In the same way the rows in the F matrix register the financing of the expenditure of institutional sectors through transfer incomes and/or credits from other institutional sectors. The relations between an institutional sector's credit or transfers from other sectors and its own total current liquid resources can exhibit stability because of institutional conditions are similar to the technical stability in the relations between the functional sectors' expenditure on intermediary goods and their gross products.

1.2.2 The stocks in the transactions table

The opening assets and liabilities of the institutional sectors are introduced in the transactions table in Fig. 1.10, page 23. The two matrices for assets, tangible assets K_{-1} and financial assets H_{-1}, are located above the B matrix. The opening liabilities are registered to the left of the Y matrix. The opening liabilities are divided betweeen debts B, equities E and wealth W.

The financial output, which contains net capital growth, will lead to changes in capital stocks, and so the closing stock values are different from the opening values. In the transactions table the closing assets are placed in the matrices K_t and H_t at the bottom of the figure and the closing liabilities can be found on the right-hand side of the transactions table in the matrix containing B_t, E_t and W_t.

However, there can be changes in the values of closing stocks compared with opening stocks without any capital growth if there are changes in the prices of tangible and financial assets during the period. Changes in the asset prices because of such price changes are registered in the V^A matrix included in the transactions table above the K_t and H_t matrices for closing assets. It should be noted that assets are recorded in market values, while the institutions which issued the securities will account for them among their liabilities in their nominal values. Therefore the valuation of securities from the creditor and the debtor sides may not be equal. However, there is a matrix among the liabilities where the price changes in liabilities are registered. It is denoted V^L.

(a) *The real and financial input–output structure*

The character of the structural relations in the transactions table can be penetrated further by concentrating on the two matrices for intermediary goods and financial transactions. The first is treated in the traditional way in input–output analysis. It is assumed that in the short run the technical conditions are constant so that the need for intermediary goods in order to produce a given gross product is constant. Therefore a coefficient matrix is introduced in the A matrix, where the coefficient is defined as the ratio of the expenditure in vector j for intermediary goods bought from vector i to the gross product of vector j:

$$a_{ij} = \frac{A_{ij}}{X_j}$$

As the gross products of the production sectors are generally assumed to be known, it is possible to calculate the production values of the sectors. The value of the production of finished goods is then given as follows:

$$B = X - AX = IX - AX = (I - A)X$$

where B is the matrix for finished goods, X is the column vector for gross products, A is the matrix of the input–output coefficient and I is the identity matrix.

The input–output system can also be used to study the factor income distribution or the development of gross profits. It is assumed that both the production sector's costs for imports of goods and services and its costs for labour services are a constant share of their total input value. The operating gross surplus will be the part of the gross product value which is left after the costs for intermediary goods, imports and labour costs are covered:

$$\Pi = X' - A'X' - ZX' - WX'$$

where Π is the vector for the gross operating surplus, Z is the vector for the import costs coefficient and W is the vector for wage cost coefficients. The prime indicates the transpose of the matrix.

Fig. 1.5 shows the transactions table including the coefficient matrices.

The F matrix in the transactions table can now be transformed into a coefficient matrix which shows how the expenditure by institutional sectors for finished goods, transfer expenditure and lending has been financed by borrowing or by transfer incomes, i.e. by 'external finance'. The ratio of total external finance to the value of total current liquid resources gives indirectly the role of the gross saving of the sectors for financing the forms of resource allocation mentioned above. This means that the disposition of total current liquid resources must be given, as well as its distribution between expenditures for finished products and for transfers and lending for every institutional sector. If this is known, it is possible to use the coefficient for financing to calculate how the sectors will cover that part of expenditure and lending which is not financed by their proper gross saving.

This calculation can also be analysed more formally:

$$R = (I - F)^{-1} Y$$

where R is the row vector of the current liquid resources of the institutional sectors, I is the identity matrix, F is the financial coefficient matrix, where the elements are the ratios F_{ij}/R_j of a sector's borrowing and transfer income to its total current liquid resources, and Y is a vector for the institutional sectors' factor income including imports of goods and services, i.e.

$$Y = \begin{bmatrix} Z \\ \Pi_1 \\ \vdots \\ \Pi_n \\ W \end{bmatrix}$$

Here Z is the total imports of goods and services, Π_i is the gross operating surplus in sector i and W is the total household wage income.

The analysis of the transactions table in input–output terms is based on the applicability of the Leontief method, i.e. output is determined by inputs. In the financial input–output analysis this means that real and financial capital growth together with expenditure for finished goods, transfers and lending are determined by financial inputs in the form of disposable income, transfer income and borrowing in the institutional sectors.

As Barker (1977, p. 98) notes, the methodology underlying the original Leontief specification is that there is an exact technical relationship between current inputs and current outputs. This seems reasonable in a single plant of a process industry, where the flow of production is not interrupted. However, in the economy as a whole, where the production sectors cover many types of industries and services which are highly heterogeneous, we can no longer regard a particular set of inputs as being contemporaneously transformed into a set of outputs. The outputs may represent an earlier set of inputs, and changes in work in progress may completely disrupt the fixed relationship. As Barker (1977, p. 99) remarks, 'It is more accurate to regard the inputs as detemined by a set of *expected* outputs which may in turn be dependent on historical outputs.' This means that we have turned the technical relation upside down. We are now maintaining that instead of inputs determining the output, current inputs are determined by expected future outputs. Therefore our analysis has been changed on two fundamental points: instead of technical relations we are working with behavioural relations, and we also assume that time enters into the analysis so that we must have a dynamic approach.

Historical relationships, where the current planned inputs may be determined by different relevant variables, enter the behavioural relationships. The measurement of such relationships may involve complicated econometric techniques.

(b) *Dynamic analysis and equilibrium*

It is possible to introduce time into an analysis working with transactions tables if we enter a transactions table for each period (Ekstedt and Westberg, 1987). The stocks of assets and liabilities built up over time can be illustrated by pillars sustaining the transactions tables. The shorter we make the periods the thinner will be the transactions tables (Fernholm, 1963; Westberg, 1983; Ekstedt and Westberg, 1987).

In the first case let us assume that the production growth rate over time from $t-1$ to t_0 is constant and that this experience during past time makes people expect a constant growth rate in the future not only in the short run to $t+1$ but further into the future (as is seen in Fig. 1.6). In a model of neoclassical shape this would mean that the actual growth rate Y was equal to the growth of capacity Y^* so that the economy would maintain full employment during the whole trajectory.

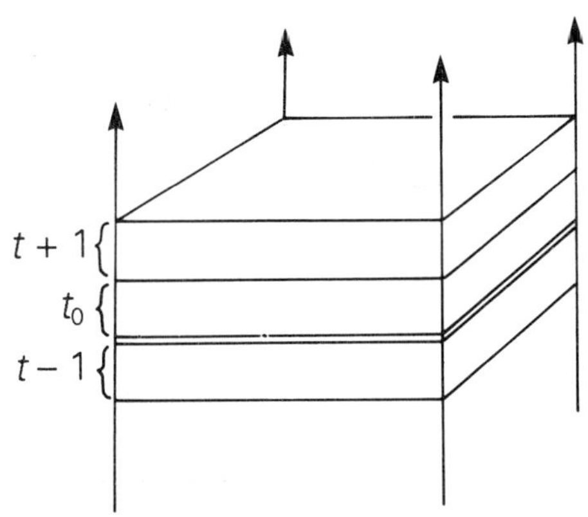

Fig. 1.6 Illustration of the transactions table in the time dimension.

In the transactions table such a development would mean that the growth rate in all markets and for all types of variables is equal. The relations are characterized by complete proportionality (as illustrated in Fig. 1.7), and this also applies to relations which are valid over time as in many behavioural relations. The result is that in such a steady state development technical relations and behavioural relations would give the same results.

If the current growth rate is not equal to the capacity growth there have to be differences between the results obtained using technical relations and behavioural relations, where stocks and flows during different time periods are related. In such a development the technical relations in the input–output method are not sufficient. The situation can be illustrated for an economy where there is a constant full capacity growth and the current growth exhibits cyclical variations (Fig. 1.8).

In the period t_0 the economy has exhibited a strong boom from the period $t - 1$. This experience influences the expectations not only for the next period $t + 1$ but also further in the future up to period $t + 2$. As the current growth rate is higher than the capacity growth rate and the capacity utilization is above the normal level, there will be a strong propensity to increase investment based on the expectation of high income growth in $t + 1$ and $t + 2$.

In order to understand the causes of the recession in period $t + 1$, it is necessary to study the changes in the financial development which take place between $t - 1$, t_0 and $t + 1$. The profit growth between $t - 1$ and t_0 decreases between t_0 and $t + 1$. The lack of labour leads to strong increases in wage rates and in the price level. The result may be increasing prices on the bond and equity markets which lead to a decrease in the demand for finished goods.

It should be noted that the short-term financial changes which lead to fluctuations in the production growth rate produce disturbances in the

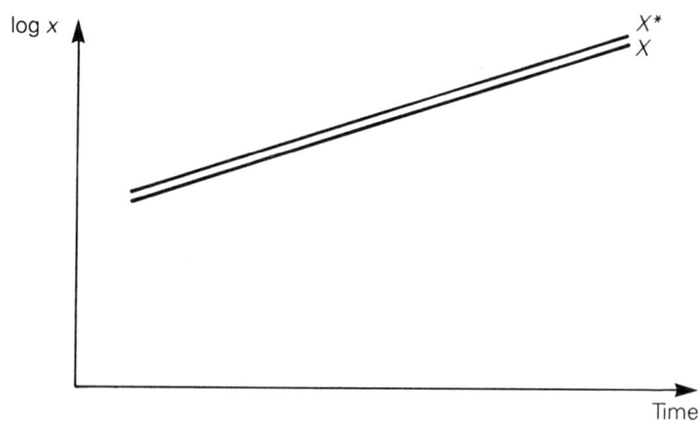

Fig. 1.7 Neoclassical steady state growth.

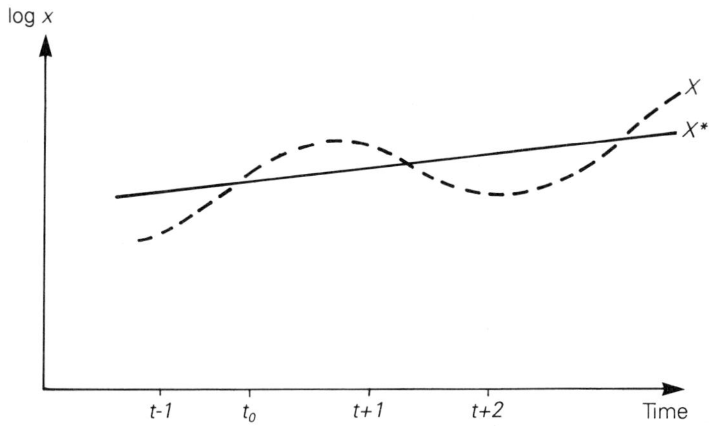

Fig. 1.8 Relation between current growth and capacity growth.

long-term rate of growth. These changes can be studied in the transactions tables for the different periods in Fig. 1.9. In particular, the relations between current expenditure on investment and consumption on the expectations of future income growth, which may depend on the historical growth rate, can be studied in Fig. 1.9.

It is clear that these dynamic relations cannot be treated using the technical equations of the real and the financial input–output systems. The input–output systems must also incorporate behavioural relations explaining the demand for finished products and for financial investments.

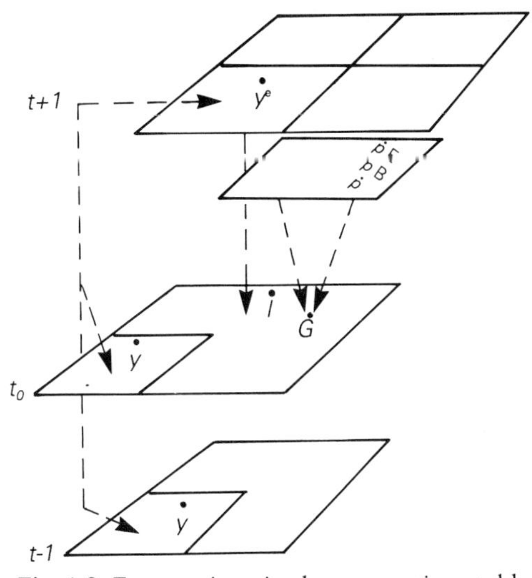

Fig. 1.9 Expectations in the transactions table.

(c) *The structure of the macroeconomic theory with a detailed exposition of the finance matrix*

The input–output analysis has shown that in a dynamic model, where there is not the same constant growth rate for all variables over the trajectory, we need macroeconomic theory for the expenditure on finished goods and on investments in financial assets.

The transactions table in Fig. 1.10 can help to illustrate the structure of this theory. The number of columns in the matrices for finished goods (the B matrix) and for financial transactions (the F matrix) is determined by the number of types of finished goods and financial assets relevant for the macroeconomic theory. The first column shows the foreign sector FOR. The foreign sector's expenditure for finished products is registered as exports X in the foreign sector's column in the row vector for finished products. The second sector includes all the domestic production sectors which belong to the institutional sector non-financial enterprises (NON). Its expenditure on finished products is gross investment I. The following three sectors represent financial enterprises: the central bank (CEN), other banks (BAN) and other credit institutes (OCI). They are assumed to have no expenditure on finished products. The final two columns represent the government sector (GOV) and the household sector (HOU). Government expenditure G for goods and services and household consumption expenditure C respectively are registered in their columns in the vector for finished goods.

The current liquid resources of the various vectors are registered in their respective rows in the column vector for current liquid resources R. The foreign sector's resources R_F are given in the first row, followed by non-financial enterprises (R_N), the central bank (R_C), other banks (R_B), other credit institutes (R_O), the government sector (R_G) and the household sector (R_H).

In addition to expenditure on finished products, which is given in the row vector for finished goods, the current financial resources are used for financial transactions, which can have two forms: transfers to other sectors or financial investments. Transfers are denoted T. Financial investment is designated as a change in a financial asset. Both transfers and financial investments are registered in the column for non-financial enterprises in the finance matrix. Starting from the bottom, the first to be registered is T_{HN} which signifies transfers from non-financial enterprises to households. These could be interest and dividends. The non-financial enterprises make both transfers T_{GN} to and financial investments dB_{GN} in the government sector. The transfers are generally taxes and the financial investments are purchases of government bonds. They also make transfers T_{ON} to other credit institutes which include payments of employees' contributions to pension funds. Non-financial enterprises pay transfers T_{BN} to banks which include interest on bank loans. The enterprises also make financial investments

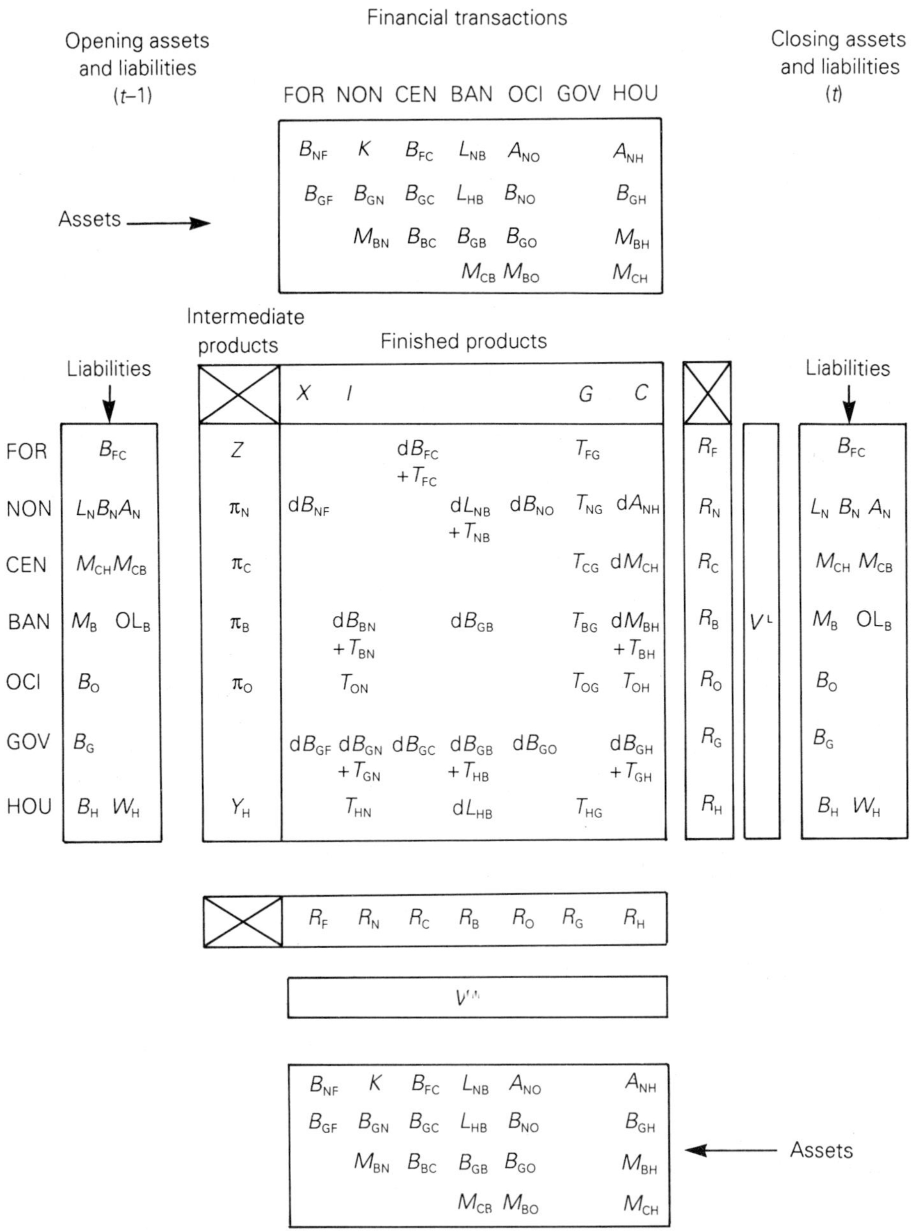

Fig. 1.10 Detailed representation of the finance matrix.

dM_{BN} in the banks which show up as changes in their bank deposits.

The financial investments in a sector show how it has changed its asset portfolio during the period. Thus the households in our example made financial investments in claims against the government (dB_{GH}), against the banks (dM_{BH}) and against the central bank (dM_{CH}), which are changes in the households' holdings of banknotes, and finally investment in equities of the non-financial enterprises (dA_{NH}).

Investments in real and financial assets lead to changes in the asset values at the end of the period compared with those at the beginning. However, the closing asset values can also be different from the opening values because of asset price changes. Such changes are registered in the row vector for asset price changes (V^{A}). If no price changes have occurred during the period, changes between the closing values at the bottom of the transactions table and the opening values at the top of the table can only be the result of investments during the period which are registered in the finance matrix (financial investments) and in the vector for finished products (real investments).

The output of current financial resources is created by inputs in the form of factor incomes, transfer incomes and borrowing. In the column vector for factor income in the transactions table the factor income of each sector is given in the row for the sector. The first row shows the income Z accruing to the foreign sector from the economy's import expenditure. The factor incomes of the non-financial enterprises shows a gross current surplus π. Indirect taxes minus subvention are registered in the government sector's row. This value is included in the value of the gross national product (GNP) calculated in market prices. Finally, the household sector's factor income Y_{H} includes wages and households' enterprise income.

In order to represent the current liquid resources of the sectors transfer income and borrowing, including new issues of equities, have to be added to the sectors' factor incomes. Therefore, for example, the current liquid resources of the non-financial enterprises are the sum of factor income π_{N}, borrowing dB_{NF} from the foreign sector, borrowing dL_{NB} from the banks, borrowing dB_{NO} from other credit institutes, subventions T_{NG} from the government and liquid resources dA_{NH} from newly issued equities bought by households.

Borrowing and new equity issues lead to changes in sector liabilities. Such changes can also be the result of price changes. However, it should be noted that debts and equities are valued in nominal values. In fact, an indication of disequilibrium is given by changes in asset valuation in the V^{A} row vector which leads to different valuations in the portfolios of the owners and among the liabilities of the issuing enterprises.

The development of government expenditure for final goods can be assumed to be dependent on the aims of financial policy. This may result in dependence on income development in the private sector and cyclical changes in capacity utilization. Exports will also be influenced by exogenous conditions as well as the relation between foreign and domestic development which changes the competitive position of the export industry. Therefore the central behavioural relations concern private consumption and gross investment by the non-financial sector. This theory will not be discussed here as it will be treated in a later chapter.

Transactions which can be regarded as exogenous are also included in the financial matrix. Among them are the majority of income transfers. Some of these, such as interest and dividends, are determined by incomes and stock values from an earlier period. The government sector transfers to and from other sectors may also in part be regarded as exogenous, depending on political decisions. Although tax rates may be decided in Parliament, taxes are endogenous if they are calculated on current incomes. An important part of household saving directed towards pension funds and insurance companies can be treated in a similar way if it is determined by fees and charges from earlier periods but calculated on the sector's current incomes.

In this way some types of financial investments in sectors with a positive financial saving can be channelled almost automatically, and the flow of funds on the credit market is governed by the distribution of factor income, which in turn is dependent in the short run on the growth rate of the GNP.

Some very important behavioural relations concerning financial investments still remain. These include, in particular, the choice between addition to liquid assets in the form of bank deposits and investment in the bond and equity market by households and other credit institutes. The banks are primarily concerned with the choice between liquid assets and loans to the public while the business sector can choose between investments in real capital and financial assets.

The transactions tables can help to illustrate how demand for both finished goods and financial investment depend on the current liquid resources of the institutional sectors in the forms of factor incomes, transfer incomes and borrowing. This means that changes in the factor income distribution—both its short-run cyclical variations and its medium-term and long-run changes—will have important effects on the demand for finished goods and financial investments. In order to cover this the demand functions for investment should include capacity utilization as well as more long-term expectations about capital yield and financial resources. The consumption function should also include variables measuring the creditworthiness of households as well as the more short-term development of disposable income.

(d) *Discrete and continuous dynamic models*

The transactions tables shown in Fig. 1.11 show that the relations between different economic variables can be clarified by dividing time into periods

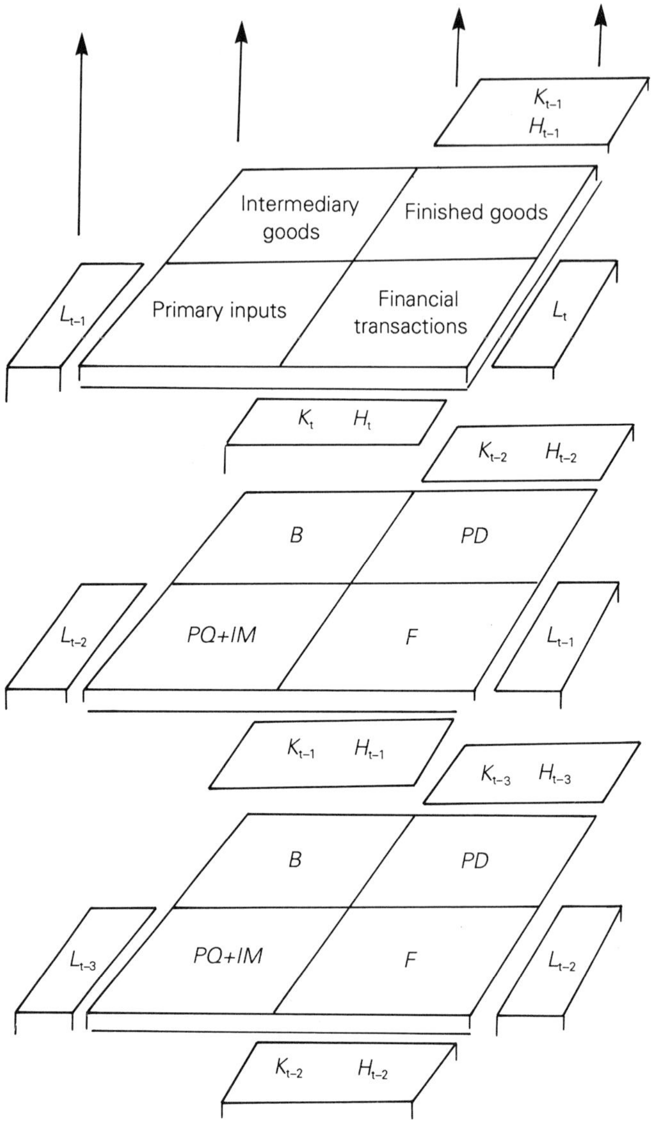

Fig. 1.11 Transactions tables with assets and liabilities for three periods.

$t-1$ representing the past, t_0 representing the present, i.e. the period which is just about to begin, and periods $t+1$, $t+2$ etc. representing the future. Such a division can also be realistic from the point of view of the economic information which will be available to agents in various sectors: 'He cannot continuously register all that happens around him' (Lindahl, 1939).

Institutional factors may also make it realistic to apply a discrete dynamic analysis if plans are changed on the basis of the results registered at the boundary points between periods, i.e. between two transactions tables as shown in Fig. 1.10 between $t - 1$ and t_0.

Therefore, for example, governments are obliged to present income and expenditure accounts before the beginning of the period which covers the budget year. If t_0 is the budget year in our example, the budget would be presented at the end of $t - 1$. The budget accounts are not only running accounts for the year $t - 1$ which is closed at the time that the account is presented. There are also forward accounts of the expenditure planned and the revenue expected for the future year t_0. Therefore government budget rules are well adapted for a discrete dynamic analysis as they give us both the *ex ante* plans and the *ex post* result of the budget policy (Hicks, 1956).

However, agents in the business or household sectors may not have sufficient control over their future operations to be able to present formal forward accounts. This means that the possibilities of *ex ante* budgeting would be of minor importance in these sectors (Hicks, 1956). Instead business plans are reflected in a statement of assets and liabilities. The valuation of the existing stock of real capital and the network of claims and obligations that are built upon it are determined by the expectations of the agents about the future development on different markets.

The expected price changes in the market for assets and liabilities may be quite different. Machinery and other real capital may be traded on markets where price movements are relatively sluggish. If stock equilibrium does not exist, producers can only restore the stocks to desired levels by changes in production.

There are other types of products, for example, certain raw materials, for which price movements may be both rapid and violent if the market is in disequilibrium. The extreme example of a market with very rapid price changes is the financial market. There is not sufficient time for the flow supply and flow demand to influence the price movements on such speculative markets. Instead, prices are determined by the liquidity preference of the dealers, i.e. by their willingness to hold stocks. This willingness is governed by their expectations about the future state of the markets.

The transactions tables also include goods and services for which the market prices can be considered to be very sluggish. This can be the case in parts of the production system where monopolistic competition dominates. However, there are also markets where the price movements may be very rapid and violent. (One can say that we follow the Keynesian pattern by assuming prices for real assets to be sluggish and prices for financial assets to be changeable.)

If prices are to be changed at the boundary points between the periods, it is obvious that it is very difficult to work with discrete dynamic models as price mobilities vary so much depending on market types. It may therefore be more practicable to use continuous dynamic models. In such models expectations are changing continuously. The modern development of data technique has made the information flows very rapid and hence a continuous analysis is more appropriate.

REFERENCES

Amoroso, L. (1937) Wilfred Pareto. *Econometrica*.

Barker, T. S. (1977) Making the Cambridge Growth Project Model dynamic: medium-term dynamic forecasting, in *The 1975 London Input–Output Conference* (ed. W. F. Goshing), Input–Output Publishing Company, London.

Bower, J. L. (1970) *Managing the Resource Allocation Process: A Study of Corporate Planning and Investment*, Irwin, Homewood, IL.

Debreu, G. (1974) Excess demand functions. *Journal of Mathematical Economics*, **1**.

Ekstedt, H. and Westberg, L. (1987) *Den finansiella utvecklingens konsekvenser för ekonomisk stabilitet. (The consequences of financial development on economic stability)* Mimeograph, Dept. of Economics, University of Göteborg.

Fernholm, T. (1963) Samhällsekonomisk orientering, Department of Economics, Göteborg, mimeo.

Hahn, F. (1985) *Money, Growth and Stability*, Basil Blackwell, Oxford.

Hicks, J. (1956) *Method of dynamic analysis*, in *Twenty five Economic Essays in honour of Erik Lindahl*, Ekonomisk Tidskrift, Stockholm.

Keynes, J. M. (1922) *Treatise on Probability*, Cambridge University Press, Cambridge.

Keynes, J. M. (1936) *General Theory of Employment, Interest and Money*, Cambridge Univesity Press, Cambridge.

Knight, F. (1921) *Risk, Uncertainty and Profit*, in *Series of reprints of scarce works on Political Economics*, Vol. 16 (1935) London School of Economics.

Leontief, W. T. (1946) Wages, profits and prices. *Quarterly Journal of Economics*, (1), November.

Lindahl, E. (1939) *Studies in the Theory of Money and Capital*, George Allen and Unwin, London.

Malkiel, B. G. (1985) *A Random Walk Down Wall Street*, Norton, New York.

Nilsson, S. and Persson, I. (1988) *Investeringsbedömning (Investment Planning)* 3rd edn, Liber, Stockholm.

Radner, R. (1968) Market equilibrium under uncertainty. *Econometrica*, **36**, 31–58.

Radner, R. (1974) Market equilibrium and uncertainty: concepts and problems, in *Frontiers in quantitative economics, Vol II* (eds M. D. Intrilligator and D. A. Kendrick) North-Holland, Amsterdam.

Radner, R. (1982) Equilibrium under uncertainty, in *Handbook of Mathematical Economics,* (eds K. J. Arrow and M. D. Intrilligator) *Vol. II*, North-Holland, Amsterdam.

Sen, A. (1976) *Rational Fools*, mimeo, London School of Economics.

Westberg, L. (1983) Finansiering i ekonomisk teori, Department of Economics, Göteberg, mimeo.

2

Production, wages and prices

2.1 THE DEVELOPMENT OF PRODUCTION IN THE INDUSTRIALIZED WORLD

Economic development in the industrialized world was characterized by a high growth rate from the end of the Second World War up to the middle of the 1970s. During the subsequent period the growth rate has generally been much lower.

In some countries the high growth rate during the period immediately following the Second World War was a result of rebuilding after the War. For example, this was the explanation for the very high growth rates for GNP in the Federal Republic of Germany (FRG) and Italy during the 1950s. In the following decade there was an increase in economic growth in several industrial countries. Liberalization of international trade increased the growth in demand, and immigration to the industrial centres increased the growth of potential production in Western Europe.

Figure 2.1 shows the difference between GNP growth rates in industrial and developing countries. As the growth rate in the industrial countries declined the prices of the developing countries' raw material exports decreased steeply and their terms of trade deteriorated, but their import prices for industrial goods generally remained more stable. In this way the import capacity of the developing countries was restricted, which caused severe hardship for the population in countries with a very low income level.

The high growth rate of the industrial countries is characterized by a higher growth rate for industrial exports and investment than the growth rate in the rest of the economy. Therefore the high growth rate is the result of both a rapid increase in foreign demand and a self-generated expansion in internal demand. The best example of this has been the post-war development in Japan.

The final phase of high growth rate in the industrial countries before the trend break in the middle of the 1970s was strongly influenced by the war in Vietnam. Increasing military expenditure was added to growing expenditure for social services during the last part of the 1960s in the United

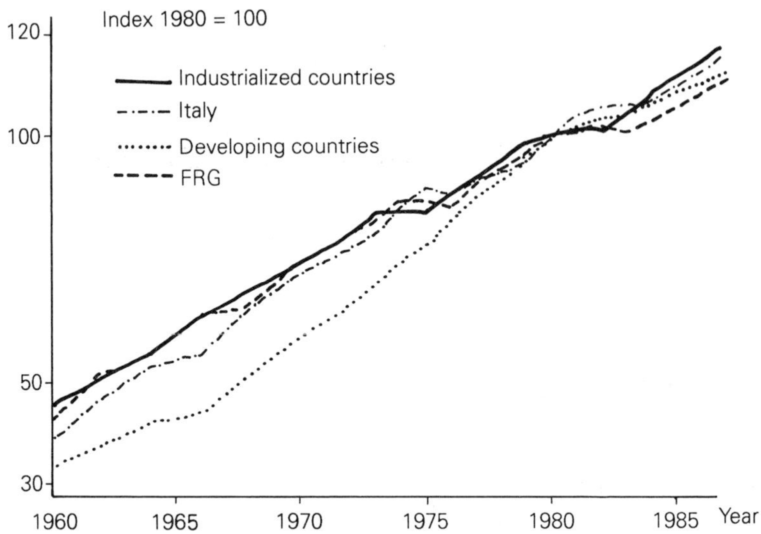

Fig. 2.1 Development of the GNP for industrial and developing countries and for some individual countries.

States. The result was an expanding growth rate and a growing deficit in the balance of payments. This development undermined the Bretton Woods system of fixed exchange rates between the leading currencies. After the *de facto* devaluation of the dollar in 1971 the countries producing raw materials and petrol products contracted in dollars experienced a deterioration of their financial position. In subsequent years this deterioration was reversed in countries producing petrol and corn. Crop failures led to increasing corn prices and the war in the Middle East made it possible for the oil-producing countries to adopt a monopolistic pricing policy, which led to oil price increases of more than 300% in a few months in 1973–4. The result was rising import expenditure in the countries without oil production, and the huge balance-of-payments deficits forced these countries to adopt restrictive economic policies which slowed down the growth rate in most industrial countries.

The disequilibrium in both industrial and developing countries was temporarily covered by the incomes of the oil-producing countries, which generally could not be invested in the home countries because of lack of investment opportunities. Instead they were transferred by the international banking system to the deficit countries. The gap between the growth rates in the industrial countries and the developing countries was reduced as the result of a somewhat higher rate in developing countries and a lower growth rate in industrial countries.

The situation changed, however, after the second oil price crisis. A very restrictive monetary policy was introduced by the Federal Reserve System in the United States in 1979. This led to a decrease in the growth rate and created a payments crisis in the developing countries as interest payments increased and new loans from the credit institutes were restricted. The most acute crisis started in Mexico in 1982, and it forced the US authorities to modify their policy. The US policy was further changed in 1985, when agreements with the other leading industrial countries led to falling interest rates and a depreciation in the dollar. The capital import necessary to cover the gap in the balance of payments and in the federal budget was maintained with the help of the rise in the equity prices, stimulated by decreasing interest rates and the expansion of private wealth, a result of the federal budget deficit.

The Western European industrial countries were more severely affected than other regions by the oil price crises. During the years 1975–9 the GNP growth rate was lower in the European OECD countries than in the OECD as a whole. The same difference in growth rates was noticeable in most years during the 1980s up to 1988. The growth rate of consumer prices has also been higher in the European OECD countries than in the OECD as a whole. Among the major powers in Western Europe both oil-producing and non-oil-producing countries are included among those with particularly high consumer price inflation (for example, the United Kingdom and Italy).

The current balances for goods, services and transfer payments have shown a better balance in the European OECD countries than in the OECD as a whole. The balances have shown a surplus since 1970 except for the 4 years 1974–7 following the first oil price crisis and the 4 years 1979–82 following the second oil price crisis. For the OECD countries taken together the deficit after the second oil price crisis has been maintained in all years up to 1988. This development is principally due to the gradually increasing deficit in the United States during the period 1982–7. In Europe, the FRG has had an increasing surplus during the same period.

After a relatively long period of stable production growth, a low inflation rate and a moderate rate of unemployment, the development in the United States began to show inflationary tendencies, although oil prices were still going down and interest rates were rising to some extent. The result was a tendency to portfolio substitution of shares by bonds and a lower demand for equities. Decreasing equity prices led to accelerated sales and market overshooting. The stock exchange crash in New York, 1987, spread all over the world and it was only because of the resolute action of the American authorities that the fall in equity prices was stopped by a huge increase in liquidity from bond purchases by the central banks. The strength of this reaction was sufficient to keep private spending on an unchanged growth rate, and the major collapse of the equity market did not influence the growth of production in the industrial world. Instead, the cyclical growth of income spread to Western Europe. This development showed that tendencies towards the high growth rate experienced during the 1960s still remained in many countries.

The new situation in the 1970s

The decreasing growth rate observed from the middle of the 1970s to the latter part of the 1980s, particularly in Western Europe, was a result of the deterioration in the terms of trade as import prices increased because of rising oil prices. Rising prices led to lower demand for finished goods, which meant lower utilization of production capacity and thus higher unit cost, as can be seen in Fig. 2.2. Another effect was a decrease in productivity. The result was that wage increases, demanded in order to compensate for falling real wages, led to higher price increases and demand for new wage increases. The inflation tended to accelerate. However, the lower productivity growth was also the result of a long-run change in the production structure in Western European countries.

2.2 THE PRODUCTION FUNCTION AS A TECHNICAL RELATIONSHIP

A structural change had been going on for a long time in the countries which were the first to be industrialized. These countries were characterized by a continuous decrease in employment in agriculture, forestry and other primary branches of production. The share of employment in industry had also been going down and the decrease accelerated in some industrial countries after the Second World War. Instead, the share of employment was increasing in both the private and public service sectors.

This development has resulted in a decrease in the average productivity growth in these countries, as the production sectors with the highest growth rate have been those with declining shares. As industrial production in most industrial countries has been of particular importance for exports, the

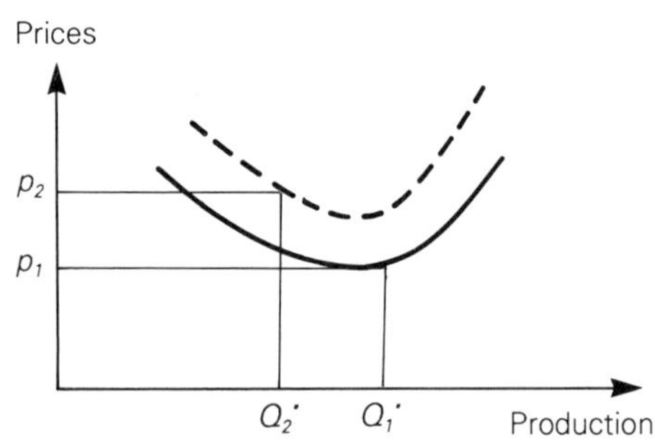

Fig. 2.2 Oil price effect on the aggregate production costs.

structural decline in industry has resulted in a declining export growth in the medium term in some countries. It has aggravated the cyclical disequilibrium that has reached severe proportions in many industrial countries during the 1970s. It has also increased inflationary effects, as productivity growth rates have declined and balance-of-payments deficits have led to devaluation and increasing import prices.

(a) *Conclusions*

We have seen that the 1950s and the 1960s were characterized by a relatively high stability, although there were some cyclical movements. During the 1970s the changes were on a different scale. There were violent decreases in growth rates and changes in income distribution.

This development has led to a growing interest in the economic theory of disequilibrium models and fixed price theory. These models analysed rigidities in the economy . However, there has been a tendency to regard these rigidities as imperfections of some type which were the result of an unsuitable economic policy. In general equilibrium models a perfect economy without any rigidities can be analysed. Such an economy reaches first-best equilibrium. If, as a result of rigidities, the economy is characterized by fixed-price equilibrium it can only reach second-best equilibrium.

Our view, however, is that the rigidities in the empirical economies are basic to the theoretical analysis because they are essential features of the economy. When we use functional relationships which *a priori* assume substitution for variables which in reality cannot be substituted, the theoretical conclusions will be false with respect to the empirical reality. We do not criticize equilibrium analysis as a tool, but we reject its use as a realistic representation of an economy.

(b) *Structural change in production*

We argued in the preceding chapter that in analysing economic disequilibrium there were three central problems which were interrelated and created problem of disequilibrium: (a) uncertainty, (b) temporary liquidity restrictions and (c) inadequate flexibility of factors of production. As we shall see in a more rigorous analysis in Chapter 3, these problems create a double role for the financial market, where the demand for liquidity as insurance for uncertainty may pervert the market's function as a mechanism for transforming financial savings into productive investments.

In this chapter we are particularly interested in the problem of inadequate flexibility and the effects of this on production costs and factor demand. We shall also show how these inflexibilities can create fluctuations in factor income distribution and thereby influence the financial flows which we study in Chapters 3 and 5.

The two basic factors of production, labour and capital, each have two roles in the economic system. They both participate in the production process and have their specific technical properties. However, labour consists of human beings, *Homo sapiens*, whose desires are the final end of the economic activity. Production capital, however, has its main role as an asset which has an expected future yield compatible with financial assets. The latter role of capital is the end, and its technical properties are the means of attaining this end. The technical properties of capital are usually specified at the time of installation and will be given during its lifetime, but as an asset it may undergo many revaluations. The two roles of capital must not be confused, but the two roles of labour are always confused.

It is impossible to separate labour's productive role from its social role, at least in a society based on liberal ideas. It is possible that individuals' preference relations include not only preferences about goods and services produced by the productive system but also the wants and desires of society as a whole and the organization of the economic system.

It is important not to confuse the two roles of capital because agents act according to capital as an asset, but its efficiency as an asset is defined by its technical properties. The Keynesian expression 'efficiency of capital' indicates the two roles and the following quotation from J. M. Keynes poses the problem:

> The absurd, though almost universal, idea that an act of individual saving is just as good for effective demand as an act of individual consumption, has been fostered by the fallacy, much more specious than the conclusion derived from it, that an increased desire to hold wealth, being much the same thing as an increased desire to hold investments, must, by increasing the demand for investments, provide a stimulus to their production; . . . It is of this fallacy that it is most difficult to disabuse men's minds. It comes from believing that the owner of wealth desires a capital-asset as such, whereas what he really desires is its prospective yield. Now prospective yield wholly depends on the expectation of future effective demand in relation to future conditions of supply. (Keynes, 1936, p. 211)

Actually the only occasion when the confusion does not matter is in general equilibrium because we then have a logical and unique connection between preferences and production factors.

In the theory of production our main intention is to analyse the technological relationships in physical production. This must be the foundation for an analysis of capital as an asset.

> It is impossible to discuss all the important theoretical developments concerning technical constraints facing the firm and its decisions under different market environments. We shall deliberately avoid discussion of market structures; both space limitations and the fact that these

> issues are discussed in other chapters of this Handbook suggest our choice. In our discussion we adopt the assumption, albeit highly unrealistic, of 'perfect competition' in both commodity and input markets. The main focus of discussion will be the description and analysis of technical constraints facing the firm, but some attention will also be given to dynamic input demand, output and price responses when regulatory constraints are present, and price and output decisions when a firm produces several products. (Nadiri, 1982)

It is clear from this quotation that the production function measures technical relationships irrespective of whether or not we are in a state of general equilibrium. Therefore, as we understand it, these technical relationships are independent of the relative valuations of inputs and outputs.

In the light of this fact we must ask ourselves whether the production functions that economists frequently use really fulfil this claim[1]. The Cambridge School obviously denies the relevance of mathematical propositions based on the neoclassical theory of production for empirical conclusions, and we shall follow and, we hope, develop their critique (Robinson, 1962). A synthesis of modern theory of production is given by Nadiri (1982) who gathers all the central axioms and assumptions.

The neoclassical production function is based in general on the existence of possibilities of aggregating inputs and outputs. It is also based on the asssumption that factors are substitutes in the production process. When these conditions hold, we are able to make assumptions about the usual properties of differentiability, positive marginal products, convex production spaces etc. Therefore we must examine the conditions for aggregation and, as a consequence, the possibilities of technical substitution of factors.

Let us start with a common production function of the Cobb–Douglas type with Hicks-neutral technical growth. We could equally well have developed the arguments around a constant elasticity of substitution (CES) function or a Leontief function:

$$Q = \exp(\gamma t) K^{\alpha} L^{\beta} \tag{2.1}$$

where Q is the production in some index, K is the capital in some index, L is the labour in some index and γ is the technical growth factor. We can use

[1] Wittgenstein analyses in *Tractatus Logico-Philosophicus* the relations between common language and logical language. Particularly in social sciences, when we use mathematics as an analytical tool the following paragraphs should always be present in our minds:

6.11 A proposition of mathematics does not express a thought.

6.211 Indeed in real life a mathematical proposition is never what we want. Rather we make use of mathematical propositions only in inferences from propositions that do not belong to mathematics to others that likewise do not belong to mathematics. (In philosophy the question 'What do we actually use this word or this proposition for?' repeatedly leads to valuable insights.)

this function to estimate long- or short-term production functions. To be meaningful, however, the functional relationship has to be stable during some period of time. Of course the functional relationship given by (2.1) will measure any combination of labour and capital, and this relationship may vary from time to time and between firms, but from the theoretical point of view we have to assume stability over time, especially with respect to a rigid capital stock. Ekstedt and Westberg (1988b) studied the Swedish industry sector in order to derive an expression for the long-term capacity growth to which short-term capacity variations could be related. We used variations in energy consumption and labour utilization relative to the long-term medium values as indicators of capacity utilization. We shall not discuss our specific use of the estimation results but use the estimates as illustrations of the theoretical problems of production functions. Therefore we have enlarged the analysis to estimates for 8 year periods between 1963 and 1986.

We used the following method for the estimates. There is a rather stable long-term relationship between capital stock and the consumption of electrical energy in the industrial sector, which we used to correct for variations in the degree of utilization of capital:

$$K^* = \frac{E}{\bar{E}} K \tag{2.2}$$

where $\bar{E} = a_1 + a_2 K$, K^* is the capital stock utilized and $\bar{E}$ is the average long-term energy consumption. Estimates for the period 1961–83 gave the following statistics:

$$a_1 = 66.74 \quad t = 12.3$$
$$R^2 = 0.75$$
$$a_2 = 0.48 \quad t = 8.9$$

According to this we constructed the production function with respect to an average capacity utilization:

$$Q_t^* = \exp(\lambda t) K_t^{*\alpha} L_t^{\beta} \tag{2.3}$$

We estimated this function in a double logarithmic form for two periods, 1961–84 and 1970–84, with no restrictions on the parameters. We obtained the results shown in Table 2.1. As we can see there are constant

Table 2.1

Period	α	β	λ	*Intercept*	R^2
1961–84	0.508	0.492	0.012	−0.68	
1970–84	0.340	0.660	−0.003	0.08	0.94
t values	(3.1)	(7.3)	(−2.8)	(0.2)	

returns to scale in both periods but the parameters are different. It is also interesting to note the radical reduction in the growth rate. However, we obtained these estimates by applying the same technique to 8 year periods from 1963 to 1986. The results are summarized in Table 2.2. We used the autoregressive estimation technique to obtain the production function. We can compare the results shown in Table 2.2 with the estimations of the differential, in Table 2.3

$$\mathrm{D}Q = b'_0 + b'_1\mathrm{D}L + b'_2\mathrm{D}K + b'_3 t \tag{2.4}$$

The changes in the parameters of the production function are very similar to the changes in the parameters in the estimates of the differentials. However, it is obvious that Tables 2.2 and 2.3 illustrate successive changes in the composition of the Swedish industrial sector. The parameter changes do not indicate a change in technology but a change in the composition of technologies.

In the quotation given above, Nadiri avoids discussion of market structures. What we try to measure using production functions is technology in an engineering sense. In studies of aggregates this is possible only under very strict assumptions of constant relations between sectors. The danger of relying on microeconomic methods in studies of aggregates is best illustrated by a comparison of Tables 2.1 and 2.2. The elegant expressions in Table 2.1 are estimated using exactly the same material and the same methods. Of course, we can draw conclusions from the estimates in Table 2.1 but hardly any of these concern production functions in a technical sense. We shall return to these conclusions later.

We use several methods of aggregating individual production functions. First, we bring individual finished goods together into one sector good. Second, we aggregate several types of capital, labour etc. into average capital, labour etc. Third, we aggregate different technologies into an average technology. These steps are obvious and most economists agree on the difficulties, but we shall discuss the techniques more closely.

Aggregation implies a search for an index function which includes some weighting of the items included in the aggregate. Technically we search for an index function defined on an n-dimensional space, the goods space for example, which is non-negative. The index function has to be strictly increasing, continuous and quasi-concave. The first two conditions imply that infinitesimally small changes in one of the dimensions of the space will cause a change in the index in the same direction. The third condition implies that the change in the index must not be greater than the change in the quantity of the goods in question. However, before we are able to construct the index function we must guarantee the existence of a measure.[2]

[2]We define a finitely additive measure according to Hewitt and Stromberg (1965) as a function μ defined on $\mathcal{A}$ such that: i) $0 \leq \mu(A) \leq \infty$ for all $A \in \mathcal{A}$. ii) $\mu(\phi) = 0$ and iii) $\mu(A \cup B) = \mu(A) + \mu(B)$ if $A, B \in \mathcal{A}$ and $A \cap B = \phi$.

Table 2.2

Period	a_0	a_1	R^2	b_0	b_1	b_2	b_3	R^2
1963–70	3.300	0.013	0.94	1.46	0.602	0.006	0.057	0.99
	(29.8)	(11.0)		(0.8)	(2.3)	(0.03)	(10.2)	
1965–72	3.370	0.011	0.93	2.01	0.468	0.050	0.051	0.99
	(28.9)	(9.6)		(1.7)	(2.3)	(0.4)	(10.7)	
1967–74	3.66	0.009	0.94	2.22	0.397	0.100	0.045	0.99
	(37.3)	(10.8)		(1.5)	(1.6)	(0.6)	(10.2)	
1969–76	4.14	0.005	0.79	2.14	0.215	0.330	0.031	0.97
	(34.0)	(5.4)		(0.7)	(0.5)	(1.3)	(4.2)	
1971–74	4.50	0.002	0.33	−0.75	0.689	0.509	0.022	0.93
	(28.8)	(2.1)		(−0.7)	(3.4)	(2.9)	(4.7)	
1973–80	4.75	0.001	0.08	−3.47	0.680	1.119	0.024	0.87
	(51.9)	(1.3)		(−1.6)	(2.3)	(3.6)	(2.9)	
1975–82	4.68	0.001	0.23	−2.57	0.232	1.378	0.008	0.87
	(44.0)	(1.8)		(−1.2)	(0.6)	(4.6)	(0.8)	
1977–84	4.11	0.002	0.49	2.47	−0.096	0.600	0.008	0.77
	(14.5)	(2.8)		(0.9)	(−0.1)	(1.7)	(0.4)	
1979–86	3.27	0.010	0.64	1.34	0.160	0.605	0.014	0.99
	(7.1)	(3.7)		(1.7)	(0.6)	(6.4)	(3.3)	

The relation between energy and capital is given by

$$\ln E = a_0 + a_1 K$$

and the production function is given by

$$\ln Q = b_0 + b_1 \ln L + b_2 \ln K^* + b_3 t$$

t values are given in parentheses.

Table 2.3

Period	b'_0	b'_1	b'_2	b'_3	R^2
1963–70	0.069 (3.8)	0.439 (1.5)	−0.076 (−0.6)	−0.003 (−0.9)	0.04
1965–72	0.050 (3.0)	0.357 (−0.6)	−0.111 (1.7)	−0.001 (−0.3)	0.04
1967–74	0.063 (9.4)	0.402 (4.3)	−0.048 (−0.7)	−0.003 (−3.1)	0.78
1969–76	0.067 (8.0)	0.539 (3.8)	0.160 (1.9)	−0.006 (−4.1)	0.78
1971–78	0.023 (0.6)	0.366 (0.7)	0.488 (1.6)	−0.001 (−0.14)	0.29
1973–80	0.023 (0.9)	0.537 (1.2)	0.877 (2.7)	−0.0004 (−0.08)	0.56
1975–82	−0.012 (−0.4)	0.168 (0.3)	1.187 (3.4)	0.003 (0.7)	0.66
1977–84	0.075 (2.1)	0.524 (1.1)	0.758 (3.1)	−0.010 (−1.9)	0.66
1979–86	0.018 (0.7)	0.426 (0.9)	0.490 (3.2)	0.0001 (0.03)	0.76

It is obvious that indices based on weight or cubic measures are possible, although the sense of aggregating a barrel of oil with a barrel of milk could be questioned. If the measure of production is numerical the problem of aggregation is clear. One car and one typewriter are hardly comparable. The continuity condition will be violated in most cases of aggregation of this kind. These considerations are elementary and rather trivial, but they are important when we leave the bright world of general equilibrium for the obscure world of disequilibrium.

In economic studies the natural way of performing aggregations is in value terms. This implies that a change in an index has two dimensions—changes in quantities and changes in the relative valuation of the relevant objects. When we base our analysis on the assumption of general equilibrium, a change in relative prices, *ceteris paribus*, is an unambiguous sign of a change in the distribution of preferences and will cause reallocations of the productive resources but will not necessarily cause any change in the index of production. However, we must then assume perfectly flexible factors of production. A change in index under the assumption of a constant distribution of purchasing power and a constant distribution of preferences indicates a change in the volume of production, and thereby a growth in real income.

In equilibrium the two-dimensional character of the index does not cause any problems because we then have a logical relationship between preferences and prices where the preference is the basic norm for the valuation of the factors of production. Observe that in general equilibrium the price formation concerns goods represented in at least one agent's preference relation. The prices of factors of production are based on the prices of finished goods under the assumption of optimal production techniques.

In disequilibrium, however, the normative role of preferences does not hold; the price formation is based on false prices relative to the distribution of purchasing power and preferences. To perform an aggregation we must then assume fixed relative prices to be able to use an index as an index of volume. When relative prices are not stable an aggregate production function will not be independent of the market structure and as a consequence it will not give precise information about the technical properties of substitution between different kinds of inputs.

The problem of substitution becomes even more crucial when we add the problem of aggregating different types of capital and labour into some average capital or labour. In the case of capital this is particularly irritating because of the large variations between different capital items with respect to flexibility among different production processes. The production capital also changes over time. We then refer to vintages of capital. Different vintages can be aggregated into a 'jelly' provided that we are in a state of equilibrium.

The problem of aggregation in disequilibrium makes it very difficult to use the production functions obtained to derive factor demand. The relation between the marginal products and factor prices will not hold in disequilibrium. In fact the marginalistic approach will fail when the economy is not in general equilibrium.

In disequilibrium, or fixed-price equilibrium, the prices are normally set according to production costs and above the price at which all the production would have been sold. An undesired stock, which in the present time period has a market price equal to zero, will then arise (Keynes, 1930). How long the producer insists on maintaining this price is a matter of competition, expectations and liquidity. In this case the goods in question will be transformed into an investment good. In an economy with sticky prices the profit will fluctuate relative to the fluctuations in demand if the factors of production are rigid. The likelihood that the capital owners will pass these fluctuations on to the labour force is mainly dependent on the flexibility of the labour force with respect to the volume of production, as we shall analyse further in this chapter.

In disequilibrium the traditional neoclassical production theory (see, for example, Malinvaud, 1977; Benassy, 1986)

$$\max \mathrm{PI} = PQ - wL - v\mathrm{K} \tag{2.5}$$

subject to

$$Q = f(K, L)$$

will be transformed to (the circumflex indicates expected values)

$$\max \widehat{\mathrm{PI}} = P\hat{Q} - w\hat{L} - vK \tag{2.6}$$

subject to

$$\hat{Q} = f(K, \hat{L})$$
$$Q \leq \hat{Q}$$

PI is profit, P is market price, Q is quantity of production, w is wage, L is quantity of labour, v is capital cost and K is quantity of capital, and $Q = g(L)$ is the actual production. If we transform the two formulations of the problem into labour demand, we have

$$\max \mathrm{PI} = PQ - wL \tag{2.7}$$

subject to

$$Q = h(L)$$

provided that the capital stock is fixed in both problems during the period of analysis. Solving this with respect to prices and wages, we obtain the labour demand

$$L^{\mathrm{d}} = h'^{-1}\left(\frac{w}{p}\right) \tag{2.8}$$

The corresponding disequilibrium model is

$$\max \mathrm{PI} = PQ - wL \tag{2.9}$$

subject to

$$Q = g(L) \qquad \text{and} \qquad Q \leq \hat{Q}$$

In this case the labour demand will be

$$L^{\mathrm{d}} = \min\left[h'^{-1}\left(\frac{w}{p}\right), g^{-1}(Q)\right] \tag{2.10}$$

When actual production is equal to the expected or potential production, we have the usual conditions with respect to prices and wages, but when actual demand is less than expected, the labour demand will be independent of the relation between prices and wages and related only to the effective demand for goods. The disequilibrium analysis implies that we impose a further restriction on the optimization problem which is independent of the relation between prices and wages. Theoretically this restriction is implied if the following condition is fulfilled:

$$g^{-1}(Q) > h^{-1}(Q) \qquad \text{if } Q < \hat{Q} \tag{2.11}$$

The most common example of disequilibrium is the difference between planned and actual investments. A change in expectations will cause a decrease in the demand for capital goods irrespective of the current price relations. The quotation from Keynes (1936) cited earlier in this chapter indicates this disequilibrium. Capital plays a crucial role because basically it is an asset that is demanded for its expected future yield and not for its own sake. In its capacity as an asset it can be substituted by unproductive financial assets. Arrow and Hahn express the problem as follows:

> In the previous chapter (13) where we assumed that money was required for every act of exchange, we found it necessary to distinguish between target and active excess demands. The distinction is quite similar to the Keynesian distinction between desired and effective demand. Strictly speaking, this distinction arises also in a barter economy: A household may have to acquire some good before it can exchange it for the desired one. (Arrow and Hahn, p. 363)

The most troublesome of all such goods is production capital, mainly because of its duration and lack of flexibility.

2.3 PRODUCTION FUNCTIONS WHEN THERE ARE COMPLEMENTARITIES BETWEEN LABOUR AND CAPITAL

In earlier discussion of economic development after the Second World War we noticed that many empirical studies have found that there seems to be a positive relationship between labour productivity and capacity utilization (Kaldor, 1956; Mazier, 1978; Dreze, 1987; Ekstedt and Westberg, 1988b; Faxén *et al.*, 1989). An explanation of this could be labour hoarding. Dreze (1987) discusses labour hoarding as rational behaviour of the firm during certain circumstances. In addition to the relationship between volume of production and labour productivity there are systematic variations in the profit share (Mazier, 1982; Ekstedt and Westberg, 1988b; Faxén *et al.*, 1989). The variations in profit share are normally explained by rigid wages which are caused by perversion of market price formation by institutional restrictions. Implicitly these rigidities cause unemployment and stagflation.

However, Dreze's discussion of the rationality of labour hoarding directs our attention to the problem of substitution between factors of production. This is also the starting point of the Leontief analysis and activity analysis. To some extent we accept labour hoarding as rational when demand is fluctuating and costs of labour training are high, but we claim that the most important reason for this is complementarity between inputs. Rader (1968) discusses this problem.

There are three main types of functional relationship between labour and production. First, there is labour which is technically tied to a specific capital equipment or some other input irrespective of the volume of production, within certain limits. Second, there is labour attached to the level of production in the ordinary technical way. Third, there is labour dimensioned to a 'normal' level of production. Examples of this last type of labour are administrative staff, sales organizations and, in particular, research departments. To these we can add institutional restrictions of different kinds, trade unions, labour legislation etc. We do not intend to analyse the effects of these latter restrictions, but it is obvious that they reduce the flexibility of production. However, it is interesting to note the enormous number of studies analysing the effects of trade unions upon the flexibility of production compared with the number analysing technical conditions which is our purpose here.

If we examine an ordinary investment decision, the capital equipment is normally dimensioned with respect to some expected optimal level of production. In this case it might be possible in the short run to use a functional relationship between labour and capital as a proxy for the third type of labour. Firms normally build up their capital equipment and develop their production organization over a period of time. To clarify our argument we assume that an investment is made at a specific time and that production is organized at the same time. It is evident that we can easily recast the problem to cover the case when capital equipment is built up over several periods and has different technical characteristics.

Let us start from an arbitrary well-specified good and a given volume of capital equipment with well-specified technical characteristics. Following the above discussion we specify two types of labour: one type is related to the volume of production and the other is dependent, directly or indirectly, on the volume of capital. Given that Q and L are non-negative, we define the following labour demand, given the capital stock:

$$L^{d} = L_0 + f(Q) \tag{2.12}$$

where f is a strictly increasing continuous function and L_0 is the volume of labour which is independent of variations of the volume of production around the target production Q^*. We assume that the labour which is independent of production in the short run is a function of the capital equipment K^* which is invested at t_0. The volume of K^* is determined by the expected prices and wages at t_0 and also by the expectation of the level of demand. In other words, we assume that at the time of investment the producers try to find the optimal organization of production which maximizes profit given the expectations. This is an important assumption, because economists usually accept that the investment decision is determined by expectations, but when there are complementarities between labour and capital we have to broaden the decision to include the choice of production organization. If we refer to the *putty–clay* approach, this includes not only the capital but also the production organization associated with the target production Q^* which maximizes profits given the expectations at the time of investment. *Putty* refers to the fact that *ex ante* the technology can be formed in many different ways; *Clay* means that *ex post* the technology is not possible to change in the short run. In many studies the concepts are used only in relation to production capital, while originally they were meant to include the total production organization.

We can illustrate the relationship between labour and production as shown in Fig. 2.3. The labour productivity z^* represents the maximum labour productivity which is attained in the short run at K^*, the invested volume of capital. The L^* is the profit-maximizing volume of labour given the expectation of future prices at t_0. Note that we discuss labour productivity in physical terms in quantities in a given unit of measure, and not in value terms.[3] The level of production at maximum labour productivity is designated Q^* and the corresponding volume of labour L^*. The ratio of capital to labour at Q^* is K^*/Q^*. Given the technology, we can assume constant returns to scale if the plants can be duplicated. The labour productivity z^* is then tied to the ratio K^*/L^*. Note that K^* is specified in time, which is important when we have technolgical change.

The curve $z = Q(L)$ represents the relationship between labour and production given the volume of capital. It then defines labour productivity

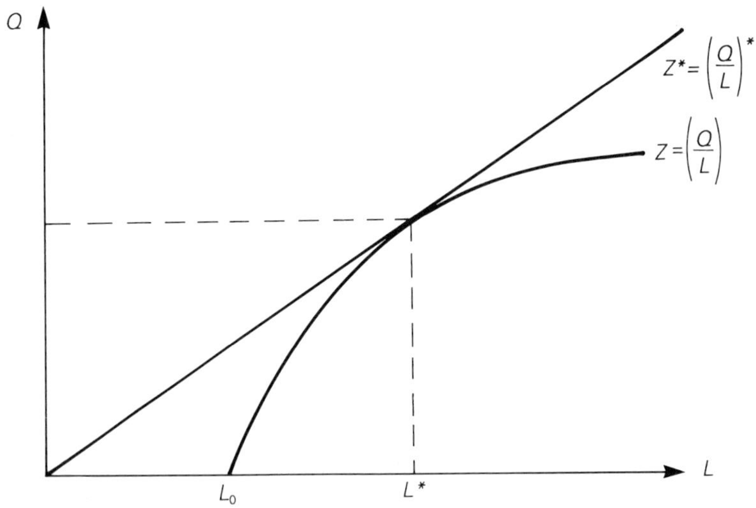

Fig. 2.3 *Ex ante–ex post* production functions at a given technology.

[3] This is not the same as production at fixed prices if the price index is an average index for a basket of goods.

at different levels of production in the short run. The conditions on the labour demand function $f(Q)$ imply that $Q(L)$ is strictly increasing and continuous when $L > L_0$. If f is twice differentiable and $f' > 0$ and $f'' < 0$ we can write

$$Q(L) = f^{-1}(L - L_0) \tag{2.13}$$

where $L > L_0$ implies $f^{-1\prime} > 0$ and $f^{-1\prime\prime} < 0$. The production function given by (2.13) is strictly concave for $L > L_0$. We can rewrite (2.13) as follows:

$$Q = g(L) \begin{cases} g(L) = 0 \text{ when } L \leq L_0 \\ g(L) > 0 \text{ when } L > L_0 \end{cases} \tag{2.14}$$

It follows from the above assumption that g is a continuous increasing function when $L > L_0$. Furthermore, it follows that it is concave in the interval $[L_0, \infty]$.

We define the average labour productivity as follows:

$$z = \frac{g(L)}{L} \tag{2.15}$$

From the definition of L^* and the conditions on g it follows that z is strictly increasing in the interval $[L_0, L^*]$ and decreasing for $L > L^*$. Differentiation with respect to labour gives

$$z' = \frac{g'L - g}{L^2} \tag{2.16}$$

We than have $g' = g/L$, i.e. $z' = z$ in z^*, which implies that the standard neoclassical marginal conditions hold *ex ante*. Equation (2.16) implies further that, under the given assumptions, $g' > z$ when $L_0 < L < L^*$ and $g' < z$ when $L > L^*$.

Before we proceed with the analysis of this production function, let us compare its properties with those of the most widely used type of production function, i.e. the Cobb–Douglas function. For a given capital stock we can write

$$Q = k_0 L^a \qquad 0 < a < 1 \tag{2.17}$$

The average labour productivity is

$$Z_{\mathrm{CD}} = k_0 L^{a-1} \tag{2.18}$$

Differentiating with respect to labour we obtain

$$z'_{\mathrm{CD}} = k_0(a - 1)L^{a-2} \tag{2.19}$$

The marginal labour productivity is

$$\frac{\mathrm{d}Q}{\mathrm{d}L} = ak_0 L^{a-1} \tag{2.20}$$

Under the assumption that $a < 1$, equations (2.18) and (2.20) state that marginal labour productivity is always less than average labour productivity.

The fundamental problem with theoretically constructed production functions is that the short-run *ex post* behaviour is derived from the long-run production function. Let us return to our original production function and derive the cost functions.

If we assume a constant wage, an assumption which we shall drop later, the total cost function is

$$\mathrm{TC} = wL^* + vK + bQ \tag{2.21a}$$

where bQ is the cost of other inputs which are assumed to have a constant relation with Q. We can transform the total cost function into

$$\mathrm{TC} = \frac{w}{z} Q + vK^* + bQ \tag{2.21b}$$

The corresponding average and marginal cost functions are

$$\mathrm{AC} = \frac{w}{z} + \frac{vK^*}{Q} + b \tag{2.22}$$

and

$$\mathrm{MC} = \frac{wz - wQ(\mathrm{d}z/\mathrm{d}Q)}{z^2} + b \tag{2.23}$$

Let us define the elasticity of labour productivity with respect to the quantity of production as $s = (\mathrm{d}z/\mathrm{d}Q)Q/z$.

In our discussion of the relations between labour and capital and the relation between goods demand and labour demand, we disregard the input costs bQ, even though they are important, particularly in an international context.

We can simplify expression (2.23) as follows:

$$\mathrm{MC} = \frac{w}{z}(1 - s) \tag{2.24}$$

It follows from the basic assumptions that s is positive in the interval $0 < Q < Q^*$ and negative for $Q^* < Q$. Furthermore, s takes values between 0 and 1 when production is below the *ex ante* optimum. In general we can conclude that the closer L_0 is to L^* in Fig. 2.4 the closer s will approach unity when labour utilization is in the neighbourhood of L_0. In fact, the maximum value of s will be a continuous function of the relation

L_0/L^*. As L_0 approaches zero, s will also become zero and $L_0 \rightarrow L^* \Rightarrow s \rightarrow 1$.[4] When $Q = Q^*$, MC will be a minimum:

$$\text{MC}^* = w/z^* \tag{2.25}$$

In the interval $Q > Q^*$ the elasticity will decrease from zero to $-\infty$ as production increases. We shall then have increasing returns when $Q < Q^*$, constant returns when Q is in the neighbourhood of Q^* and diminishing labour productivity when $Q > Q^*$.

Before proceeding further, let us illustrate our results diagrammatically. The labour productivity curves are shown in Fig. 2.4 and the cost curves in Fig. 2.5. We do not claim that there is any novelty in these curves, but we believe that it is important to show that the shapes are natural when there are complementarities between capital and labour and capital is a fixed factor during some period of time.

A rather general picture which we teach our students (although we do not generally relate it to a mathematical function) is shown in Fig. 2.6 (see, for example, Varian, 1987). We call the interval $Q < Q^*$ the economic region, and in this interval the rational producer will be under perfect competition. Our attitude is that *ex ante* when investors plan the production organization, given expected prices, wages and the growth of demand, their target production is Q^*. But capital is a fixed factor, which means that when the investment is undertaken the market value is the present value of its expected future yield. If then, *ex post*, it appears that expectations *ex ante* were wrong, the investor will experience a capital loss, which Keynes calls a windfall loss. The investor must now choose between liquidating this loss or continuing to produce at a lower rate of return on the production capital. This means that it might be rational to produce even in the so-called non-economic region.

[4] Defining

$$s = \frac{\partial z/z}{\delta Q/Q}$$

we set

$$\frac{\Delta z}{z} = \left(\frac{Q + \Delta Q}{L + \Delta L} - \frac{Q}{L}\right) \Big/ \frac{Q}{L}$$

If L_0 is close to L^* we expect large production increases from small additions to the labour employment. We therefore require to determine the limit when ΔL approaches zero:

$$\lim_{\Delta L \rightarrow 0} \left\{\frac{\Delta Q L - Q \Delta L}{L(L + \Delta L)}\right\} \Big/ \frac{Q}{L} = \frac{\Delta Q}{Q}$$

That is

$$\frac{\Delta z}{z} = \frac{\Delta Q}{Q} \Rightarrow s = 1$$

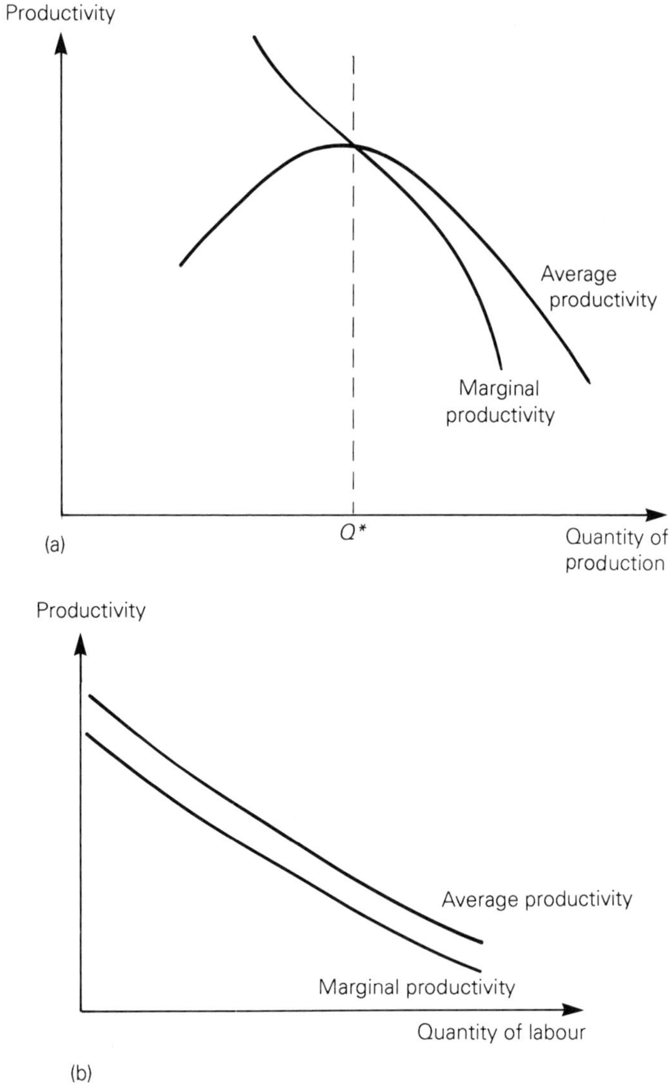

Fig. 2.4 (a) Labour productivity curves when there are complementarities; (b) corresponding curves in the substitution case (Cobb–Douglas).

These results are important in disequilibrium theory, when investment decisions are irreversible. Starting from the *ex ante* expectations of demand, we find that if the *ex post* demand is higher than the *ex ante* expectation, we shall have rising marginal costs in production, but it also seems probable

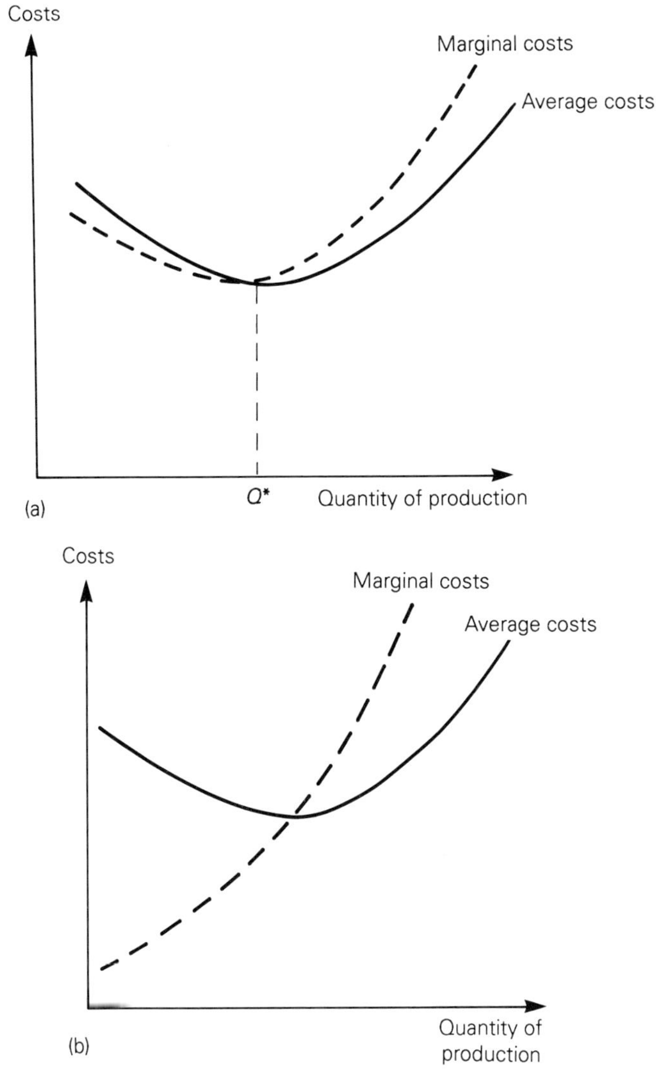

Fig. 2.5 (a) Cost curves when there are complementarities (labour and capital costs only); (b) cost curves in the substitution case (Cobb–Douglas).

that the excess demand will result in a price rise. It is interesting to ask whether prices or wages are most flexible. Alternatively, if we face a lower demand *ex post* than expected *ex ante*, we shall also face rising marginal costs and rising average costs. In this situation, however, the effects of price increases will be more dubious.

Taking these facts into consideration, it is natural to expect the following.

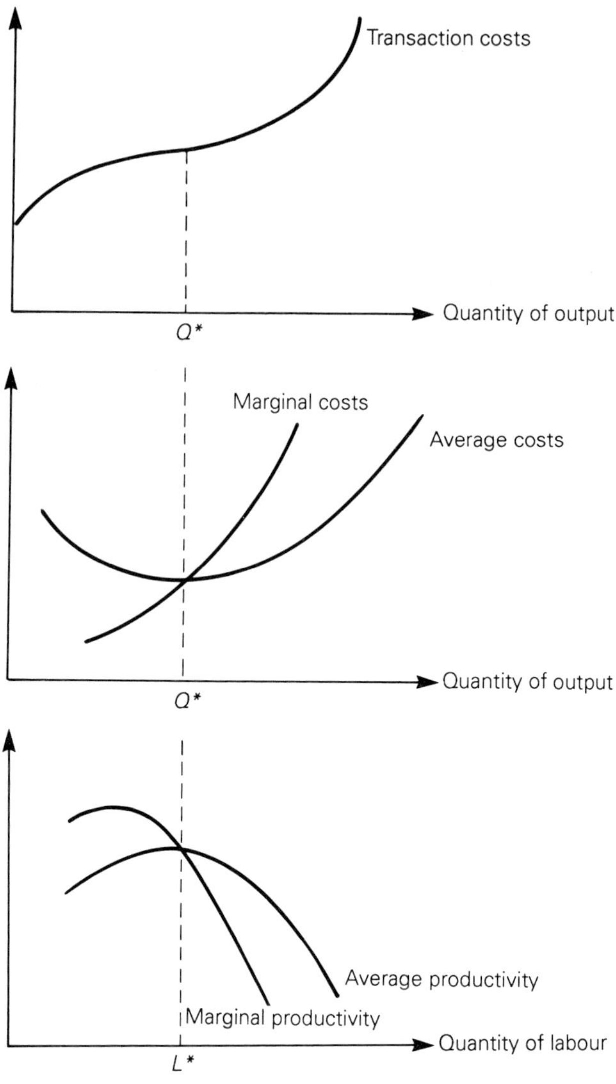

Fig. 2.6 Textbook picture of relations between productivities and costs.

1. When undesired stocks occur, the labour demand is inelastic with respect to the decrease in the demand for goods.
2. When undesired stocks occur it is natural to expect rather sticky prices. The degree of stickiness is dependent on the elasticity of demand and the degree of competition in the market.
3. Deficient demand will affect profits more than wages. We could then expect decreasing profit share when the demand for goods decreases.

4. We will face precyclical variations in labour productivity and a very close relationship between labour productivity and profit share. Figure 2.7 illustrates the point. The maximum increase in labour productivity will occur slightly after the economy has started to recover, and it will approach zero and eventually turn into negative values around the boom.

Before we begin the technical discussion of our approach we remind ourselves of Verdoorn's law. In 1949 P. J. Verdoorn published an article in which he claimed that there was a positive relationship between the volume of production and labour productivity. The explanation of this relationship was based on long-term processes such as labour division and embodied technical change.[5] Kaldor (1966) discusses Verdoorn's law and reformulates it into a relation between productivity growth and the growth of the labour force. He also finds, in his empirical investigations, that the law is most relevant for the secondary production sector, i.e. manufacturing industry. The original specification of Verdoorn's law is

$$\delta z = a_1 + b_1 \delta Q \tag{2.26}$$

where z is labour productivity and Q is volume of production, and the reformulation by Kaldor and by McCombie and de Ridder (1984) is

$$\delta L = -a_1 + (1 - b_1)\delta Q \tag{2.27}$$

where $\delta z = \delta Q - \delta L$ and L is the volume of labour.

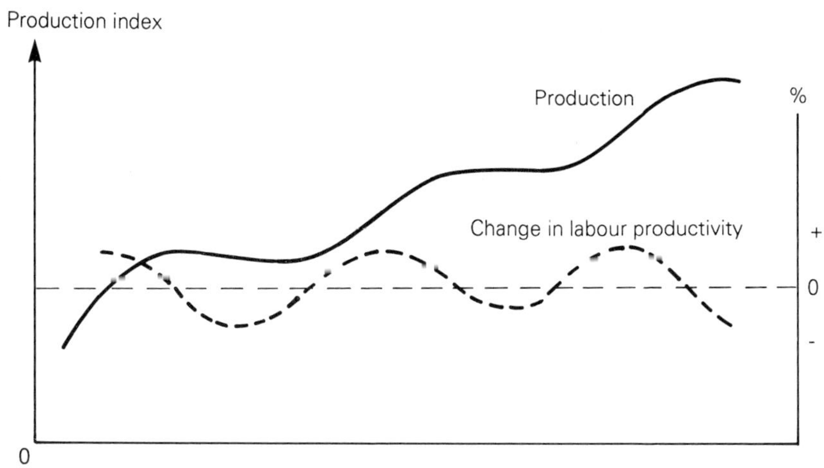

Fig. 2.7 Idealized picture of the relation between variations in production and variations in productivity.

[5]One would have expected *a priori* to find a correlation between labour productivity and output, given that the division of labour only comes about through increases in the volume of production; therefore the expansion of production creates the possibility of further rationalization which has the same effect as mechanization (Verdoorn, quoted by Boulier, 1984)

In an empirical study of American manufacturing industry, for the period 1963–73, McCombie and Ridder (1984) found strong confirmation of Kaldor's argument that the law is mainly a result of economies of scale, both internal to the firms and external in the sense of learning by doing and the total organization of the production sector. While the empirical evidence is strongly in favour of Verdoorn's law, the theoretical interpretations are not unambiguous. Most of the discussions have concentrated on the long-run factors. In general the explanations focus on economies of scale, learning by doing, the general increase in human capital and the development of social and economic organization. In a theoretical analysis of a model based on a Cobb–Douglas production function with Hicks-neutral technical growth, Boulier concludes:

> Much has been made of the findings of Verdoorn and others that the rate of growth of labour productivity is positively correlated with the rate of growth of output. However, surprisingly little attention has been paid to the theoretical relations between the aspects of economic change. Using a simple microeconomic model, I have demonstrated that the elasticity of labour productivity with respect to output depends in a complicated way on characteristics of the production function, factor supply and output demand and that differences in empirical values of this elasticity offer little insight into the nature of the underlying economic circumstances that give rise to these differences. (Boulier, 1984)

It should be clear from our preceding discussions of the relations between labour and capital and labour and production that, if there are partial complementarities between labour and capital, we would have a Verdoorn effect when the production is lower than the *ex ante* expectation and a Phillips effect if the production is higher than the *ex ante* expectation. This means that if we start from a situation with a low degree of capacity utilization and the production increases, we could expect increasing productivity in both a marginal and an average sense. However, if we start from the *ex ante* optimum production, we should expect decreasing labour productivity.

Lucas (1970) and Sargent and Wallace (1974) discuss the empirical procyclical variations of productivity and the wage share which contradict the theory of diminishing returns. They work with both Cobb–Douglas and CES functions. Lucas focuses mainly on shift work and overtime and explains variations of labour productivity and wage share mainly by introducing overtime work and differences in the capital–labour ratio between ordinary working time and overtime working. He stresses that diminishing productivity prevails in each instance, but when firms have to hire overtime workers or shift-workers for 'later' shifts they hire fewer workers per capital unit and therefore they have to pay higher real wages.

Sargent and Wallace show that the variations in wage share and in productivity are a consequence of low elasticity of substitution between labour and capital. The elasticity will be lower at 'later' shifts and consequently we shall have a relation between the elasticity of substitution and the real wage. They also show that the Lucas approach using a CES production function can be manipulated to explain complicated variations of productivity, wage share and real wages. In some sense this approach is similar to ours. The main difference is that we start from a disequilibrium context and look at profit as a residual. Furthermore, we stress the 'putty–clay' view that the technique is fixed *ex post*.

Boddy (1985), in an inquiry into US manufacturing industry and the total US business sector, shows that the hypothesis of decreasing labour productivity does not hold for manufacturing industry for the period 1955–77. However, when he considers the total private business sector, he finds decreasing labour productivity. In his investigation he uses a capacity index as a correction for the capital stock employed.

One of the central issues of Keynesian theory is the question of the stickiness of prices relative to the stickiness of wages. Let us sum up our discussions so far in this chapter using a small dynamic model of prices and wages. Let us then assume that firms use some mark up pricing. We first define the profit function as a residual between revenues and labour costs:

$$\pi = PQ - wL \tag{2.28a}$$

Inserting productivity, we obtain

$$\pi = L(Pz - w) \tag{2.28b}$$

We assume that both price changes and wage changes are functions of the excess demand:

$$\begin{aligned}
Q^{d} &= \psi_1 \frac{w}{p} L + \psi_2 \frac{\pi}{p} & \psi_1 > 0,\ \psi_2 > 0 \\
Q^{s} &= Lz \\
L^{d} &= L_0 + \phi(\mathrm{MC}) \\
L^{s} &= \bar{L} + \mu w & \mu \geq 0 \\
\dot{P} &= \gamma_1 (Q^{d} - Q^{s}) & \gamma_1 > 0 \\
\dot{w} &= \gamma_2 (L^{d} - L^{s}) & \gamma_2 > 0
\end{aligned} \tag{2.29}$$

The demand for goods consists of consumer demand, which is dependent on real wage income, and investment demand, which is dependent on real profit. The supply of goods is the short-run production function transformed to an expression in labour volume and labour productivity. Around the optimum production, labour demand is a function of production volume. Here we use an inverse relationship for the marginal costs below the optimum production and a positive relationship above. The labour supply is assumed to consist of one part which is independent of the nominal wage and one part which is dependent on it. We assume a positive dependence, although this is not self-evident in industrial countries. Empirical investigations of the Swedish labour force have shown that there is some ambiguity in the sign of the elasticity of labour supply with respect to the nominal as well as the real wage rate. The changes in prices and wages depend on the excess demand in the ordinary way. However, we assume linear excess demand functions.

For production levels below Q^* we reduce the system to

$$\begin{aligned} \dot{P} &= \gamma_1\left[\frac{w}{p}(\psi_1 - \psi_2) + z(\psi_2 - 1)\right] \\ \dot{w} &= \gamma_2\left[\frac{\phi z}{w(1 - s)} - (\bar{L} + \mu w)\right] \end{aligned} \tag{2.30a}$$

and for production levels above Q^* we have

$$\begin{aligned} \dot{P} &= \gamma_1\left[\frac{w}{p}(\psi_1 - \psi_2) + z(\psi_2 - 1)\right] \\ \dot{w} &= \gamma_2\left[\frac{\phi_2 w(1 - s)}{z} - (\bar{L} + \mu w)\right] \end{aligned} \tag{2.30b}$$

Solving the price function in equilibrium gives

$$\frac{w}{p}(\psi_1 - \psi_2) + z(\psi_2 - 1) = 0 \tag{2.31}$$

which we solve for p:

$$p = \frac{w(\psi_1 - \psi_2)}{z(1 - \psi_2)} \tag{2.32a}$$

Provided that the elasticity of consumer demand with respect to real income is in the neighbourhood of 1, which is not an unrealistic assumption, we can simplify expression (2.28) to

$$p = w/z \tag{2.32b}$$

which is familiar and for optimum production, according to 2.16, will give us the usual neoclassical result. However, it is important to note that this condition of price equilibrium states on the one hand that the wage must vary in the same direction and with the same magnitude as the labour productivity and on the other hand that price equilibrium is independent of the volume of employment.

Before we begin the analysis of the wage equation, we illustrate the output price as a function of production volume at constant wages (Fig. 2.8). If we are to the right of the optimum volume, we expect a decreasing

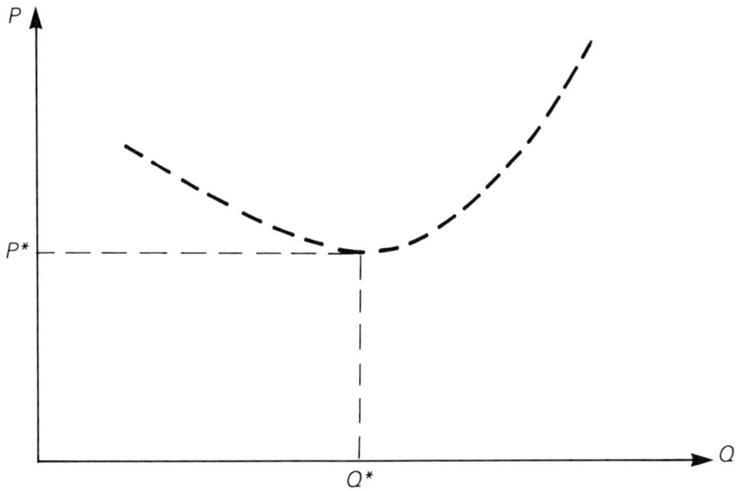

Fig. 2.8 Price–equilibrium curve.

price level when capacity utilization increases. To the left of the optimum volume we have the traditional Phillips relation. Note that we only analyse the relation between prices and wages and disregard other factors influencing the prices. The behaviour of the prices when production is below the *ex ante* optimum is dependent on the relation between L_0 and L^* in Fig. 2.3.

Let us now turn to the wage equation (2.29). To simplify the argument we can consider two cases. In the first case we omit that part of the labour supply which is inelastic with respect to wages. In the second case we assume zero elasticity of labour supply with respect to wages. For production levels below Q^* we have the following results.

Case 1 We can write (2.30a) as

$$\dot{w} = \gamma_2\left[\frac{\phi z}{w(1 - s)} - \mu w\right] \tag{2.30a'}$$

Wage equilibrium, $\dot{w} = 0$, implies

$$w = \frac{(\phi z)^{1/2}}{\mu(1 - s)} \tag{2.33}$$

Case 2 We can write (2.30a) as

$$\dot{w} = \gamma_2\left[\frac{\phi z}{w(1 - s)} - \bar{L}\right] \tag{2.30a''}$$

Wage equilibrium, $\dot{w} = 0$, implies

$$w = \frac{\phi z}{L(1 - s)} \tag{2.34}$$

For production levels above Q^* we have, analogously,

$$w = \frac{[\mu(1 - s)]^{1/2}}{\phi_2 z} \tag{2.35}$$

and

$$w = \frac{\bar{L}(1 - s)}{\phi_2 z} \tag{2.36}$$

In both (2.33) and (2.34) the wage must vary in relation to labour productivity to be in equilibrium. However, when we look at the elasticity of labour productivity with respect to production volume s, we find that the greater is the share of the total labour force which is complementary to capital stock, the more sticky the wages will be downwards in a suboptimal situtation.

Looking at the total system, which then has two variants

$$P = w/z \quad \text{transformed to} \quad P = w/p \qquad (2.37\text{a}')$$

$$w = \frac{(\phi z)^{1/2}}{\mu(1 - s)} \quad \text{transformed to} \quad w = \frac{(aZ)^{1/2}}{1 - s}$$

and

$$P = w/z \quad \text{transformed to} \quad P = w/z \qquad (2.37\text{a}'')$$

$$w = \frac{(\phi z)}{L(1 - s)} \quad \text{transformed to} \quad w = \frac{bZ}{1 - s}$$

we find that the conditions for total equilibrium are

$$P = \frac{a}{z(1 - s)} \quad \text{and} \quad w = \frac{a}{1 - s} \tag{2.38}$$

for system (2.37a′) and

$$P = \frac{b}{1 - s} \quad \text{and} \quad w = \frac{bs}{1 - s} \tag{2.39}$$

for system (2.37a″).

From (2.38), i.e. when the wage elasticity of labour supply is positive, we find that when the *ex post* production is below the *ex ante* planned level, prices have to be reduced more than wages to attain a new price–wage equilibrium. Note that such equilibria exist below the optimum production. In (2.39), i.e. when the labour supply is inelastic with respect to wages, we find that prices should be at a higher level than wages relative to the optimum relation if production is below optimum when a new equilibrium is attained.

From Equations (2.37a′) and (2.37a″) we find that the wages will be more sticky when the elasticity of labour supply is positive than when it is zero. However, although the parameters of the two expressions are different we find the same fundamental condition of equilibrium:

$$\frac{P}{w} = \frac{\bar{L}}{z} \tag{2.40}$$

We showed in (2.15) that when *ex post* production Q is equal to the *ex ante* planned value Q^*, the average productivity z will be equal to the marginal productivity. We therefore conclude that the neoclassical optimum condition of labour demand will hold when $Q = Q^*$.

When the actual production does not coincide with planned production, the marginal productivity will be replaced by the average productivity in the equilibrium condition. However, it follows from the basic assumptions of our model that the average labour productivity will attain a maximum at Q^*. When this is applied to the equilibrium condition in (2.40) it implies that relative to the first-best equilibrium, when *ex post* demand equals *ex ante* expected demand, the prices have to be higher relative to the wages in equilibria when *ex post* production/demand is below *ex ante* expected production/demand. The interpretation of this result is that a fundamental condition of equilibrium is the constancy of factor income distribution. The decreased labour productivity must be met by higher prices relative to wages in order to keep a constant profit share. The condition is also in accordance with Keynes's view that price–wage equilibria exist at lower real wages than in general equilibrium and consequently at lower demand and unemployment. The condition is also present in the proofs by Smale (1974) and Mas-Colell (1985, Chapter 6) that equilibria which are determined by the production technology exist in addition to the general equilibrium.[6]

Let us now turn to the situation in which the *ex post* demand is greater than the *ex ante* expected demand. From Equations (2.32a) and (2.36) we find the equilibrium conditions

$$P = \frac{w}{z} \qquad w = \bar{L}\,\frac{z}{\phi_2(1 - s)} \tag{2.41}$$

when the elasticity of labour supply with respect to wages is zero and provided that $\psi_1 = 1$. When the elasticity of labour supply is positive the corresponding conditions are

$$P = \frac{w}{z} \qquad \frac{\phi_2(1 - s)}{\mu} \tag{2.42}$$

[6] As a conclusion to Chapter 6, Mas-Colell states: 'Thus as f (excess demand function) satisfies (ii) the unique price vector p, and consequently the set of basic activities $A(p)$, is independent of f, it is determined by the production technology alone. This is the so called nonsubstitution theorem.'

The equilibrium conditions in (2.41) imply deteriorating nominal wages when labour productivity decreases and the same equilibrium relation between prices and wages as when $Q < Q^*$.

We find that the equilibrium condition (2.42) is independent of the nominal wage. Provided that $\phi_2 > \mu$, which is normal for most realistic production functions, we shall never obtain an equilibrium when *ex post* demand is greater than *ex ante* expected demand. This is a consequence of the change in sign of the elasticity s of labour productivity with respect to production.

The stability properties of the model are rather complicated. Rewriting the dynamic equations, we have the following variants when $Q < Q^*$:

$$\dot{P} = H(P, w) = \gamma_1 \left[\frac{w}{p} (\psi_1 - \psi_2) - z(1 - \psi_2) \right] \tag{2.43a}$$

$$\dot{w} = G_1(w) = \gamma_2 \left[\frac{\phi_1 z}{w(1 - s)} - \bar{L} \right]$$

$$\dot{P} = H(P, w) = \gamma_1 \left[\frac{w}{p} (\psi_1 - \psi_2) - z(1 - \psi_2) \right] \tag{2.43b}$$

$$\dot{w} = G_2(w) = \gamma_2 \left[\frac{\phi_1 z}{w(1 - s)} - \mu w \right]$$

When $Q > Q^*$ there is only one variant:

$$\dot{P} = H(P, w) = \gamma_1 \left[\frac{w}{p} (\psi_1 - \psi_2) - z(1 - \psi_2) \right] \tag{2.44}$$

$$\dot{w} = G_3(w) = \gamma_2 \left[\frac{w(1 - s)\phi_2}{z} - \bar{L} \right]$$

Starting with (2.43a) we differentiate the two functions H and G_1 with respect to prices and wages:

$$H'_P = -\gamma_1 \frac{w(\psi_1 - \psi_2)}{P^2}$$

$$H'_w = \gamma_1 \frac{\psi_1 - \psi_2}{P} \tag{2.45}$$

$$G'_{1,w} = -\frac{\phi_1 z(1 - s)}{w^2(1 - s)^2}$$

The essential parameters of the price function H determining the stability are the elasticities of consumption and investments with respect to real wages and real profit respectively. We can then isolate stability regimes from system (2.45) depending on whether the elasticity of investments with respect to profit is greater than or less than the elasticity of consumption with respect to the real wage

Figures 2.9 and 2.10 show the stability properties. In Fig. 2.9 the equilibrium is stable provided that the price level is equal to or above the wage level. This illustrates the Keynesian stable unemployment equilibrium. The condition is that the elasticity of consumption with respect to real wage income is greater than the elasticity of investment with respect to real profit. When the capacity utilization is low this seems to be realistic. In Fig. 2.10 the equilibrium is saddle-shaped. In the wage dimension it is stable, but in the price dimension it is unstable. The reason for this is the high elasticity of investment with respect to profit which, unlike consumption demand, is positively dependent on price.

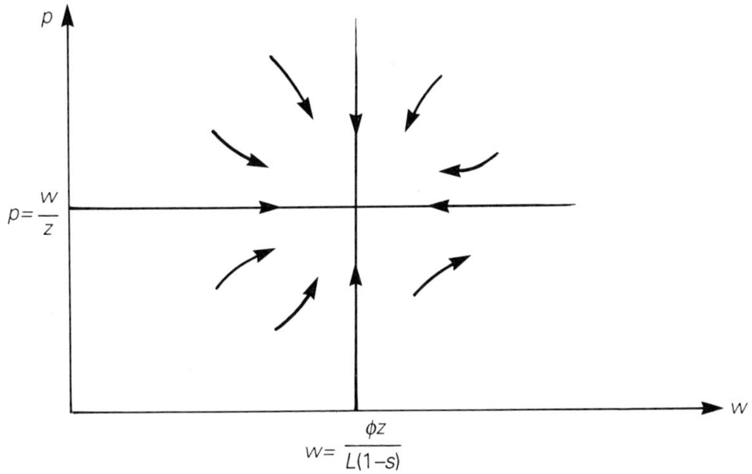

Fig. 2.9 Stability properties when $\psi_1 > \psi_2$ at a given labour productivity and $Q < Q^*$.

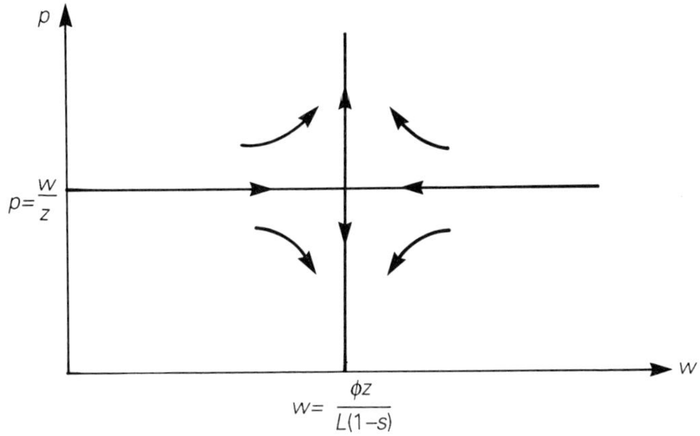

Fig. 2.10 Stability properties when $\psi_1 < \psi_2$ at a given labour productivity and $Q < Q^*$.

We assumed above that the elasticity of consumption demand with respect to the real wage income is close to unity. It also seems natural to assume rather small variations in the elasticity depending on the variations in capacity utilization. Ekstedt and Westberg (1988b) found that an increase in unemployment resulted in an increase in the rate of saving but a decrease in the ratio of saving to disposable income.

In contrast, it seems reasonable to assume that the elasticity of investment demand with respect to real profit has more pronounced variations with respect to varying regimes of capacity utilization. An increase in real profit at a low rate of capacity utilization would have almost no effect on the level of investment, while equally good prospects of profits at a high rate of capacity utilization would result in a substantial increase in the level of investment. We could even envisage a high level of investment when labour productivity, and the rate of profit, is decreasing because of a shortage of capital.

Our discussion of stability shows that if the elasticity of consumption can be assumed to be almost constant, while the elasticity of investment varies according to the level of capacity utilization, we would expect a bifurcation when the labour productivity changes. That means that the dynamic properties of the system will change from a stable to an unstable situation or vice versa. The central issue of a discussion of stability should be the behaviour of investors with respect to changes in real profit in different regimes of capacity utilization.

The stability properties of a system in which actual production is greater than planned are obtained by differentiating system (2.44) with respect to prices and wages (we disregard system (2.43b) when it has qualitatively the same properties as (2.43a)):

$$\begin{aligned} H'_p &= -\frac{Y_1 w(\psi_1 - \psi_2)}{P^2} \\ H'_w &= \frac{Y_1(\psi_1 - \psi_2)}{P} \\ G'_{3,w} &= \frac{(1 - s)\phi_2}{z} \end{aligned} \tag{2.46}$$

The fundamental change from (2.45) is that the derivative of the wage change function is positive and is decreasing with respect to labour productivity. We might then expect increased instability compared with the situation when $Q < Q^*$ and also when labour productivity decreases.

The system is unstable when the price/wage condition is fulfilled but not the condition for wage equilibrium. The equilibrium will be of saddle character but extremely sensitive to wage changes.

The stability properties are the opposite of those in Fig. 2.12. We have a saddle point here also, but the price dimension is stable when the price

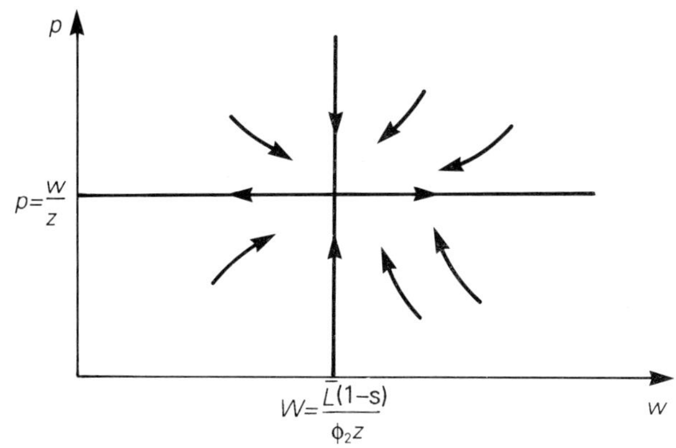

Fig. 2.11 Stability properties when $Q > Q^*$ and $\psi_1 > \psi_2$ at a given labour productivity.

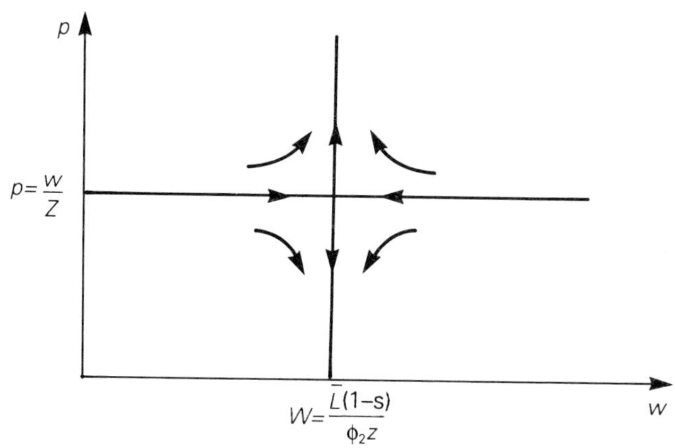

Fig. 2.12 Stability properties when $Q > Q^*$ and $\psi_1 < \psi_2$ at a given labour productivity.

wage relation is fulfilled otherwise the system is unstable. Contrary to the situation when consumer demand was more sensitive than investment demand, this situation will lead to high sensitivity to price distortions.

Comparing the two alternative stability regimes when $Q > Q^*$ it is most likely that the latter will occur more frequently because, as we pointed out earlier, the sensitivity of investment demand with respect to profit will be higher than the sensitivity of consumer demand when capacity utilization is high.

(a) *Conclusions of the short-run model*

We have argued that it is theoretically incorrect to derive a short-run production function from the long-run function just by keeping the capital stock constant. The short-run production function will behave qualitatively differently from the long-run function. In particular, the functional relationship between the volume of labour and the volume of production will be different from the long-run relationship at constant capital stock. The reason for this is that labour and capital are certainly substitutes *ex ante*, but when the investment decision is taken part of the employed labour force will not be related to the short-run variations of production volume but rather to the capital stock itself either directly or indirectly as an organization related to the optimal sales which could be represented by the volume of capital. In this respect capital and labour will be complementary factors of production.

Given these basic assumptions we have shown that equilibria exist on both sides of the *ex ante* production volume. The stability analysis has shown that stable equilibria exist when real wages and employment are below the *ex ante* values. Furthermore, we have shown that if *ex post* production is above the *ex ante* planned value, the eventual equilibria will be unstable under all circumstances. The most important parameter determining the stability properties, except those for labour productivity, is the sensitivity of investment demand with respect to real profit.

The achieved results are in accordance with both Keynesian theory (Keynes, 1936) and the theory of regular production economies developed by Smale, Mas-Colell and others.

(b) *Structural change and sector composition*

So far we have analysed a one-sector model in the short run. We must then ask in what way this sector will develop in the long run and whether the short-run stability conditions will change. We must also ask whether the introduction of another structurally different production sector will change the basic results.

Returning to Tables 2.1 and 2.2 we can see the increase in the importance of capital in Swedish industry since the 1960s. Up to the beginning of the 1970s the volume of employment increased, but since then the employed labour force has decreased steadily while the average labour productivity has increased. These results are shown in Fig. 2.13.

We must now ask whether that part of labour which is independent of short-run variations will decrease or increase relative to the *ex ante* profit-maximizing volume. It seems natural to assume that it will increase. If we consider two moments of time in Fig. 2.3 between which there has been a technical development met by an equal increase in demand for goods, thus leaving the optimum volume of labour unchanged, we obtain the diagram shown in Fig. 2.14. This kind of technological development will increase the

Fig. 2.13 Labour volume and labour productivity in Swedish manufacturing industry 1960–88 (SIND, 1986).

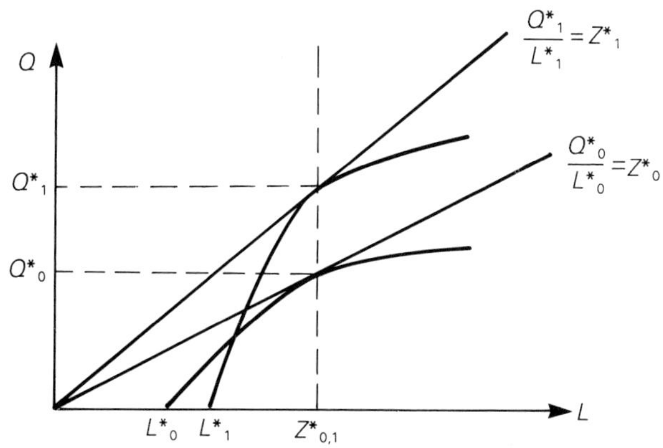

Fig. 2.14 Proportional change between demand for output and labour-saving technical growth.

variability of labour productivity. This is a consequence of the fact that a smaller part of the labour demand will depend on the variations in the volume of production in the short run. This will lead to larger variations in the profit share if *ex post* demand does not equal the *ex ante* expectation. Looking at the conditions of (2.45) and (2.46) we can see that more pronounced variations in labour productivity will in general increase stability when production is suboptimal and increase instability when actual production is greater than the *ex ante* value.

It is also clear from our analysis that if production is expanding from a low production level compared with optimum production to a higher level, we should expect higher wages, both nominal and real, decreasing prices (or at least a decreasing inflation rate) and increasing labour productivity. These effects will become more pronounced when the ratio L_0/L^* in Figs 2.3 and 2.11 increases, i.e. when that part of labour which is complementary to capital increases relative to the optimal labour volume.

Thus in the short run (within the trade cycle) we shall have a Verdoorn effect, i.e. a positive correlation between employment volume and productivity, but only when the production is below the optimum production inherent in the capital stock. Given the capital stock, we can identify two regimes, the Verdoorn regime and the Phillips regime, as shown in Fig. 2.15. The Verdoorn regime exists to the left of Q_1^* and the Phillips regime exists to the right of Q_2^*. Curve 1 represents a situation where the optimum utilization of the existing capital stock is close to full employment, while curve 2 shows the case in which the optimum degree of utilization of the capital stock is substantially below full employment.

A technical development of the kind shown in Fig. 2.14 will make the Verdoorn effect more pronounced. It will also make the Phillips curve steeper when the functional relationship between labour and production becomes weaker. If the growth of the capital stock is unsatisfactory with respect to full employment, we shall face a situation where the Phillips curve becomes steeper at higher unemployment rates.

The long-run Verdoorn effect, i.e. a positive correlation between the volume of production and labour productivity, is quite natural. An increase in capacity is caused by an increase in the capital stock which will then work at higher productivity rates. Whether or not the Kaldor (1966) version holds, i.e. that volume of employment will be positively correlated with labour productivity, depends on the growth of demand. If the growth of

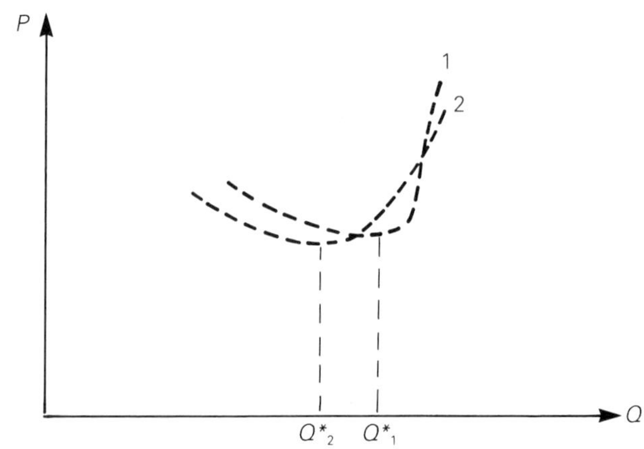

Fig. 2.15 Change in price–equilibrium curves resulting from labour-saving technical change.

demand is the same as or higher than the sum of technical growth and the growth of the capital stock it will hold; otherwise it will not.

A long-term development along the lines indicated in Fig. 2.14 will certainly have an effect on investment behaviour. As we showed in the discussion of the short-run conditions, firms will become more vulnerable when they experience variations in demand. The higher the ratio L_0/L^* in Fig. 2.14, the more severe will be the variations in profit and profit share. If firms then use, for example, the variance of rate of return on capital as a measure of risk, the risk of production capital will clearly increase. We shall return to this question in discussing investment, but an increase in the risk of production capital could make the economy move from curve 1 to curve 2 in Fig. 2.15.

(c) *The composition of production sectors*

So far we have discussed a production sector where labour and capital are substitutes *ex ante* but not *ex post*. The other type of production sector is where, given a volume of capital (buildings, equipment etc.), the production is proportional to the volume of employment. Examples of this are most kinds of health care, production of education and cultural services, restaurants etc. The central problem with these kinds of services is that in general it is difficult or impossible to increase the ratio of output in physical terms to labour hours. The productivity curves of such a production sector, given the capital stock, are generally as shown in Fig. 2.16. Until the optimum volume of employment is reached the labour productivity is constant, but after L^* is achieved it will decrease.

Baumol (1970) analyses a simple but effective model in which he includes sectors of the type we are discussing here:

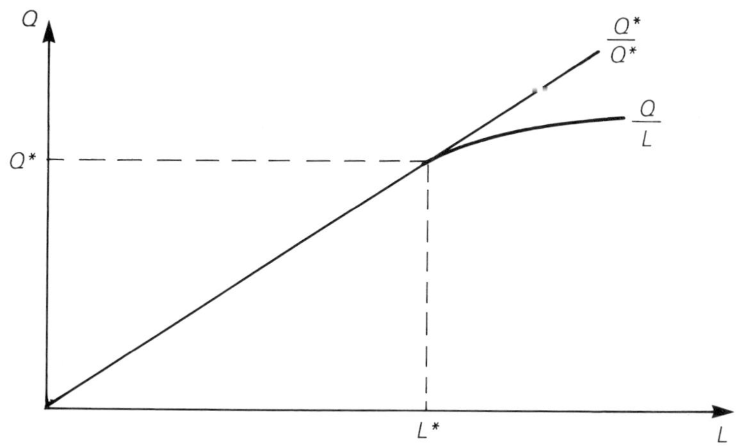

Fig. 2.16 *Ex ante–ex post* production functions in labour-intensive production.

$$Q_{1t} = aL_{1t}$$
$$Q_{2t} = bL_{2t}R^{\phi t} \quad (2.47)$$

where Q_{it} is the volume of production and L_{it} is the volume of employment. If the wages of the two sectors are growing at the same rate as the productivity increase ϕ of sector 2, he comes to the following conclusions.

1. The costs of sector 1 will increase without limit.
2. If sector 1 is competing in a market and the elasticity of demand with respect to the price is negative and non-zero, and the income elasticity does not approach infinity, the sector will vanish in the long run.
3. If, however, the proportion of the two sectors is kept constant, by for example government action, an increasing part of the labour force will be transferred to sector 1.
4. If the proportion of the two sectors is kept constant, the average productivity of the labour force will decrease.

In a closed economy the relative growth of the two sectors does not matter (as long as the government has enough political support). However, if sector 2 is competing in the international market and the country depends on imported goods, there will be a severe conflict if sector 2 faces the production conditions we have discussed earlier in this chapter. We shall return to this conflict when we discuss government policy in an internationalized world.

REFERENCES

Arrow, K. J. and Hahn, F. (1971) *General Competitive Analysis*, Holden-Day, San Francisco, CA.

Arrow, K. J. and Hurwicz, L. (1958) On the stability of the competitive equilibrium I. *Econometrica*, **26**.

Arrow, K. F., Block, H. D. and Hurwicz, L. (1959) On the stability of competitive equilibrium II. *Econometrica*, **27**.

Barro, R. J. and Grossman, H. I. (1971) A general disequilibrium model of income and employment. *American Economic Review*, **61**.

Baumol, W. J. (1970) *Economic Dynamics*, Macmillan, New York.

Benassy, J. P. (1986) *Macroeconomics: An Introduction to the Non-Walrasian Approach*, Academic Press, New York.

Boddy, R. (1985) A specious solution to the 'problem' of procyclical productivity. *Journal of Political Economy*, **93**.

Boltho, A. (ed.) (1982) *The European Economy: Growth and Crisis*, Oxford University Press, Oxford.

Boulier, B. L. (1984) What lies behind Verdoorn's law? *Oxford Economic Papers*, **36**.

Debreu, G. (1970) Economies with a finite set of equilibria. *Econometrica*, **83** (3).

Debreu, G. (1974) Excess demand functions. *Journal of Mathematical Economics*, **1**.

Dreze, J. (1987) Work sharing. *Journal of Economic Policy*, **October**.

Ekstedt, H. and Westberg, L. (1988a) A dynamic model for the interrelations of real and financial growth. Paper presented at the International Conference of Economic Modelling, London, mimeo.

Ekstedt, H. and Westberg, L. (1988b) Interaction between economic growth and financial flows, in *Economic Modelling in OECD Countries* (ed. H. Motamen), Chapman and Hall, London.

Faxén, K. O., Odhner, C. E. and Spånt, R. (1989) *Lönebildningen i 90-talets Samhällsekonomi (The Wage Formation in the Economy of the 1990s)*, Rabén and Sjögren, Simrishamn.

Grandmont, J-M. (1982) Temporary general equilibrium theory, in *Handbook of Mathematical Economics*, Vol. II (eds K. J. Arrow and M. D. Intriligator), North-Holland, Amsterdam.

Hansen, B. (1951) *A Study in the Theory of Inflation*, Allen and Unwin, London.

Hewitt, E. and Stromberg, K. (1965) *Real and Abstract Analysis*, Springer-Verlag, Heidelberg.

Hicks, J. R. (1939) *Value and Capital*, Clarendon Press, Oxford.

Hirsch, M. W. and Smale, S. (1974) *Differential Equations, Dynamical Systems and Linear Algebra*, Academic Press, New York.

Kaldor, N. (1966) *Causes of the Slow Rate of Growth of the United Kingdom. An Inaugural Lecture*, Cambridge University Press, Cambridge.

Kaldor, N. (1975) Economic growth and Verdoorn's law. A comment on Mr. Rowthorn's article. *Economic Journal*, **85**.

Kalecki, M. (1971) *Selected Essays on the Dynamics of the Capitalist Economy 1933–1970*, Cambridge University Press, Cambridge.

Keynes, J. M. (1930) *Treatise on Money*, Vols I and II, Macmillan, London.

Keynes, J. M. (1936) *General Theory of Employment, Interest and Money*, Cambridge University Press, Cambridge.

Lucas, R. E. (1970) Capacity, overtime and empirical production functions. *American Economic Review, Papers and Proceedings*, **64**.

Malinvaud, E. (1977) *The Theory of Unemployment Reconsidered*, Basil Blackwell, Oxford.

Mas-Colell, A. (1977) Regular, nonconvex economies. *Econometrica*, **45** (6) 1387–1407.

Mas-Colell, A. (1985) *The Theory of General Economic Equilibrium: A Differential Approach*, Cambridge University Press, Cambridge.

Mazier, J. (1978) *La macroéconomie appliquée*, Presses Universitaires de France, Paris.

Mazier, J. (1982) Growth and crisis—a Marxist interpretation in the European economy, in *The European Economy* (ed. A Boltho), Oxford University Press, Oxford.

McCombie, J. S. L. and De Ridder, J. R. (1984) 'The Verdoorn law controversy': some new empirical evidence using U.S. State data. *Oxford Economic Papers*, **36**.

Nadiri, M. I. (1982) Producers' theory, in *Handbook of Mathematical Economics*, Vol. II (eds K. J. Arrow and M. D. Intriligator), North-Holland, Amsterdam, Chapter 10.

Rader, T. (1968) Normally, factor inputs are never gross substitutes. *Journal of Political Economy*, **76**.

Robinson, J. (1958) *The Accumulation of Capital*, Macmillan, London.

Robinson, J. (1962) *Essays in the Theory of Economic Growth*, Macmillan, London.

Sargent, T. J. and Wallace, N. (1974) The elasticity of substitution and cyclical behaviors of productivity, wages and labour's share. *American Economic Review, Papers and Proceedings*, **64**.

SIND PM (1986) *Investeringar, Finansiering och Lösamhet i 22 Företag (Investment, Financing and Profitability in 22 Companies)*, Staten Industriverk, Stockholm.

Smale, S. (1974) Global analysis and economics IV: Finiteness and stability of equilibria with general consumption sets and production. *Journal of Mathematical Economics*, **1** (2).

Smale, S. (1982) Global analysis and economics, in *Handbook of Mathematical Economics*, Vol. II (eds K. J. Arrow and M. D. Intriligator), North-Holland, Amsterdam, Chapter 8.

Solow, R. M. (1970) *Growth Theory*, Clarendon, Oxford.

Tatom, J. A. (1980) The 'problem' of procyclical real wages and productivity. *Journal of Political Economy*, **88**.

Varian, H. (1987) *Intermediate Microeconomics*, Norton, New York.

3

Factor income distribution: short-run and long-run changes in the factor income shares of sectors

The aim of our analysis in this chapter is to show the importance of both cyclical variations in production growth and long-run structural change of production in the development of factor income distribution. The short-run variations in demand lead to cyclical movements in capacity utilization and productivity growth. When productivity growth rises during a boom, the profit share will go up and will reach its maximum shortly before the peak of the cycle. In recession, with decreasing capacity utilization and productivity growth, the wage share of the GNP will increase and will reach its maximum value shortly before the trough of the cycle. This cyclical pattern of changes in factor income distribution is fairly general in industrial countries and is comparatively easy to explain.

There are also medium- or long-term changes in income distribution. These changes are much less uniform. There have been examples of both long-term increases and decreases in the profit share in industrial countries since the end of the Second World War. These different tendencies can be explained by different types of structural changes in countries with different levels of economic maturity and different growth rates for capital accumulation.

3.1 THE DEVELOPMENT OF FACTOR INCOME DISTRIBUTION IN SOME INDUSTRIAL COUNTRIES

A calculation of the aggregate value of the share of the gross operating surplus in the seven largest industrial countries (United States, Japan, FRG, France, United Kingdom, Italy and Canada) from 1967 to 1985 shows both cyclical variations and some tendency to change in the medium term (Faxén *et al.*, 1989, p. 40). There are three maxima: in the latter part of the 1960s, in the latter part of the 1970s and after 1985. The three maxima can be explained in terms of business cycle development in the seven major OECD countries.

The GDP growth rate was at a high level in the cycle in 1968, 1976 and 1984 (OECD, 1989). However, at the beginning of the 1970s there was a decrease in the share of gross surplus that did not coincide with the cyclical development. The highest GNP/GDP growth rate for the 1970s was in the OECD countries in 1973. The share of gross surplus did not increase in this year. In the latter part of the 1970s and the early part of the 1980s the share of the gross surplus decreased to a lower level than normal during periods of stagnation. However, this can be explained by the marked decrease in demand, which was the result of a very restrictive monetary policy in the United States. After the change in US monetary policy in 1982 there were increases in the gross surplus share for each year from 1983 to 1985.

A study of the share of gross operating surplus in manufacturing in six European countries for 1955 to 1975 shows completely different behaviour (Boltho, 1982, p. 15). In this case there were also cyclical variations with maxima near the peaks on the business cycle. However, these variations were completely dominated by a strong declining trend. This decline was particularly strong during the early part of the 1960s and the first half of the 1970s. It can also be demonstrated that in a large group of European countries during the same period corporate income, excluding entrepreneurial income, was decreasing relative to national income at factor cost. This group of countries included France, the FRG, the United Kingdom, Belgium, Finland, the Netherlands, Sweden and Switzerland. The decrease was slower between 1960 and 1970, and stronger between 1970 and 1975. From 1975 to 1978 there was an increase in corporate income share.

The cyclical and medium-term changes in factor income distribution can also be studied by examining the development of the share of wages in GDP at current prices in some industrial countries in Europe (Mazier, 1982). In Italy, the FRG and the United Kingdom there are clear cyclical movements from the middle of the 1950s up to the 1970s. During the 1970s there are rising wage shares, which seem to be of a structural character, in the three countries. In France the development shows another pattern. There is a very strong decrease in the wage share during the 1950s. After 1970 the wage share increases in a similar way to the other three countries.

The wage share, calculated as the ratio of total wages and salaries to the nominal value of the GNP, has been computed for the United States, Japan, France, the FRG and the United Kingdom (Mazier, 1978). The increasing share of wage workers has been taken into consideration in these calculations, and this has resulted in a reduction of the share of wages in the GNP recorded in the national accounts.

In all countries except Japan cyclical variations in the wage share are dominant in the period between 1950 and 1975. In all countries except the United States and the United Kingdom there is an increasing trend for the wage share between 1970 and 1975. From 1955 to 1970 the trend was decreasing in France, the FRG and the United States. In contrast, in the United Kingdom there was an increasing trend during this period. In Japan the development was very special. During the 1950s there was a marked reduction in the wage share, followed by cyclical variations on an almost constant level during the 1960s. During the period 1970–75 there was a strong increase in the wage share.

The share of wages in Swedish industry showed a slightly rising trend for 1963 to 1986. However, there are some very different tendencies underlying this trend. From 1963 to 1974 the wage share decresed slightly. This was followed by a very steep increase between 1974 and 1978. From 1978 to 1983 there was a decrease of almost the same magnitude. In addition to these violent changes, which seem to be related to the oil price crises and the subsequent restrictive economic policy in the United States and other industrial countries, it is possible to note typical cyclical movements. The share of wages exhibits maxima a short time before the troughs of the business cycle, and minima shortly before the peaks of the cycle.

The data presented here seem to confirm the short-run and long-run changes in factor income distribution discussed in the introduction to this chapter. While the short-run cyclical movements seem to show similar tendencies in several industrial countries, with procyclical movements in profit shares and contracyclical movements in wage shares, the medium-term changes in factor income distribution in different industrial countries show much less concordance.

3.2 THE ROLE OF FACTOR INCOME DISTRIBUTION IN EARLY BUSINESS CYCLE THEORY

The theory used to explain these cyclical and medium-term changes in factor income distribution was primarily developed in business cycle theory and is related to the short-term cyclical changes in factor income shares. Economists such as Schumpeter (1939, p. 105), Mitchell (1950, p. 20) and later Lundberg (1968, p. 15) emphasized changes in factor income shares as an explanation of movement in expenditure and thus of business cycle fluctuations.

Changes in factor income distribution are of special importance in Mitchell's business cycle theory of fluctuations in expenditure. He notes that the increase in profits at the beginning of a boom is caused by the slow rise in supplementary costs relative to the physical volume of business. This leads to an increase in the productivity growth rate. However, investment will not expand until capacity utilization has reached a satisfactory level. Therefore maximum investment is not reached until the second or the third year of recuperation. In this phase rises are usual. The result is increasing wages and increasing family income: 'The waning, like the waxing of prosperity, therefore, must be due, not to the influence of disturbing causes from outside, but to processes that run regularly within the world of business itself'. (Mitchell, 1950, p. 26)

The boom can continue as long as prices and wages are rising at the same rate. If these relations are upset, maladjustments will occur and some enterprises will lose profits. The final phase of the boom is characterized by an increasing growth rate in wages because of labour shortages (Mitchell, 1950, p. 189). Mitchell also notes that profits shrink to a larger extent than wages during business depressions.

Although the variations in income during the business cycle must be seen as a result of changes in demand, according to Mitchell's theory the changes in factor income distribution, which in turn will influence demand for consumption and investment, can be said to be caused by the supply conditions in business (Evans, 1969, p. 340). Mitchell emphasizes that marginal costs will remain steady or even decrease when production is rising. This can be analysed as a situation which is described by the short-run Verdoorn effect.

The negative range of the supply curve in Fig. 3.1 indicates that an increase in the production level (or production growth rate) results in an increasing growth rate in productivity as costs do not increase proportionally to the growth in production. This means that during a boom, which makes the GNP increase from Q_0 to Q_1, there will be an increase in the

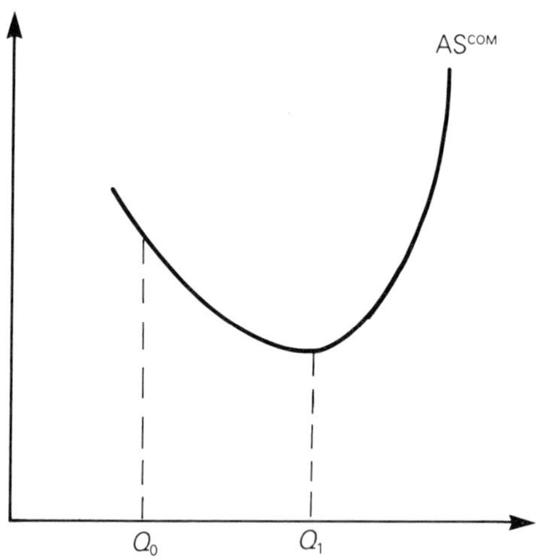

Fig. 3.1 Relation between costs and actual production in commodity production sector.

growth rate of productivity which will lead to an increasing profit share. Costs will not start to rise until the optimum capacity is reached at Q_1, when shortages of labour and input goods lead to rising marginal costs. At this production level a decrease in the profit share will lead to a decline in economic activity.

The steepness of the supply curve is dependent on the possibility of large-scale production benefits. Such benefits are greater in commodity production than in service production, which explains why the negative slope of the supply curve is steeper in commodity production than in service production. Figure 3.1 represents commodity production and Fig. 3.2 represents service production. In service sectors, where there is no international competition, there may be more room for price increases when costs increase. Therefore the service sectors may be more likely to keep up the profit share with the help of price rises when productivity growth is decreasing than the commodity sectors who work under international competition. Thus in the service sectors the Verdoorn effect may be replaced by a Phillips effect. Figure 3.2 shows that the increase in production leads to a very low increase in productivity as the supply curve is almost horizontal. Therefore the Verdoorn effect is of very little importance. An increase in production after Q_1 has been reached leads to increasing marginal costs. If it is possible to cover these cost increases with higher prices, business may be able to keep unchanged profit shares. The risk is that this Phillips effect may result in new wage increases and the cost curve will move up, as has been observed in industrial countries particularly during the 1970s (Faxén *et al.*, 1989, p.160).

The supply curve in Fig. 3.1 shows that in commodity-producing sectors there will be a decrease in productivity and in the profit share if production

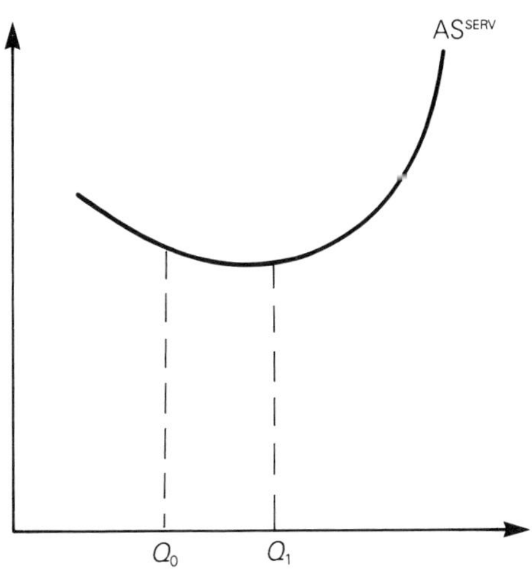

Fig. 3.2 Relation between costs and actual production in service production sector.

decreases from Q_1 to Q_0. This is the result of the industry's policy of keeping labour resources unchanged, as the increase in labour costs is believed to be smaller than the 'transactions costs' which would be necessary if labour had to be re-employed and retrained when expansion starts again.

3.3 ANALYSES OF THE ROLE OF FACTOR INCOME DISTRIBUTION IN THE BUSINESS CYCLE: KALECKI AND JOAN ROBINSON

In Kalecki's studies of economic dynamics factor income distribution played an active role in both the short and the long run (Kalecki, 1965, p. 28). In the short run that interests us here he showed how the wage share of national income changes in the business cycle. He adds salaries to wages and notes how, because of their 'overhead' character, salaries will fall less than total income during a depression and will also rise less than total income during a boom. This means that the average value of the share of wages and salaries in the gross income of the private sector will be smaller than the marginal value of this ratio.

Kalecki (1965, p. 59) gives the following relation for the share of wages and salaries in private gross income:

$$V = \alpha Y + B$$

where V represents wages and salaries, Y is private gross income, B is a positive constant for the short period, although it is subject to long-run changes, and $\alpha < 1$, as $V < Y$ and $B > 0$. If both sides of the equation are divided by Y we obtain the following equation for the share of wages and salaries in private gross income:

$$\frac{V}{Y} = \alpha + \frac{B}{Y}$$

In the short run α and B will be constant and the share of wages and salaries will fluctuate contracyclically, going up during recessions and going down during expansions. This tendency has been reinforced since the time of Kalecki's studies, as the transactions costs for training newly recruited workers are so high that businesses prefer to keep workers employed during recession.

Kalecki's analysis confirms the hypothesis that the supply curve is downward sloping in most of the income range from trough to peak. Such a situation is described by the short-term Verdoorn effect. Kalecki's stress on the overhead character of business salary costs strengthens the steepness of the downward-sloping tendency of the supply curve. The increase in productivity growth will be reinforced when production is increased during expansion. During recession productivity will go down, and businesses may have to increase prices to cover increasing costs.

Joan Robinson

The central theme in Robinsons's analysis is the reciprocal relation between the rate of profit and the rate of accumulation in the private sector of the economy (Robinson, 1962, p. 48). An increase in the rate of profit leads to an increase in the rate of accumulation and vice versa. In this way the cyclical variations in the rate of profit and thus in the factor income distribution become both a result and a cause of the business cycle fluctuations in the private sector of the economy. Robinson's constructions of the relations between the rate of profits and the rate of accumulation make it possible to analyse the condition for equilibrium and the behaviour of the economy if it is not in equilibrium.

The equation expressing the dependence of the rate of profit on the rate of accumulation is assumed to be linear, while the rate of accumulation is a non-linear function of the rate of profit. The model will thus include two equilibria as shown in Fig. 3.3. The A curve shows the expected rate of profit on investment as a function of the rate of accumulation. The I curve shows the rate of accumulation as a function of the rate of profit that induces it. The form of the I curve shows the decreasing elasticity of investment with repect to the expected profit rate. In this way the model shows that there is a maximum rate of accumulation even at a very high rate of expected profit. However, with the help of multiples, changes in the

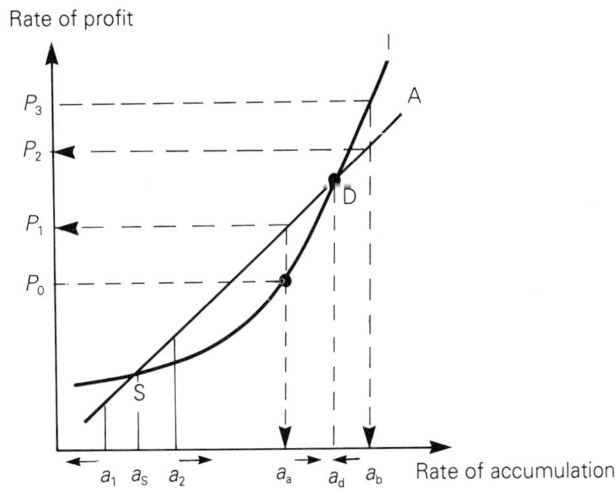

Fig. 3.3 Relations between rate of profit and rate of accumulation according to Robinson.

rate of accumulation will create proportional changes in income and profits.

Figure 3.3 indicates that there are two equilibria S and D. At D the economy has reached a stable equilibrium. If the economy has reached a higher rate of accumulation than the one which prevails at D, the profit rate induced by this rate of accumulation will not be sufficient to maintain it. The A curve in Fig. 3.3 shows that an accumulation rate of a_b creates a profit rate p_2. However, the I curve shows that in order to maintain an accumulation rate of a_b, the expected profit rate has to be p_3. The result will be a decrease in the rate of accumulation until equilibrium D is reached at accumulation rate a_d.

If the accumulation rate were lower than a_d, the profit rate induced by this accumulation rate would be higher than the expected profit rate necessary to maintain it. This is the situation at the accumulation rate a_a, where the induced profit rate is p_1 but the expected profit rate is only p_0, and it explains why the accumulation rate tends to grow as long as the induced profit is greater than the expected profit. It is not until the accumulation rate a_d is reached that it will be stabilized, as the induced and expected profit rates are equal when equilibrium D is stable.

The other equilibrium in the model is unstable, however. This equilibrium is indicated by S. If the accumulation rate is lower than a_s, as it is at a_1, the induced rate of profit is lower than the expected rate. Thus the accumulation rate will move away from the equilibrium rate a_s. If the accumulation rate is greater than a_s, as for instance at a_2, the A curve is above the I curve. This means that the induced rate of profit is above the expected rate, and hence there is an increase in the accumulation rate. This increase will continue until the equilibrium at D is reached, as has been seen before. This means that the equilibrium at S is unstable.

It has been shown that Robinson's condition for equilibrium is that the induced and expected rates of profit are equal. In disequilibrium the expected accumulation rate is either higher or lower than the rate of profit induced by this rate of accumulation. Such states of disequilibrium characterize the various phases of the business cycle. In the expansion phase there is a high rate of accumulation, which tends to increase as the induced rate of profits is higher than the expected rate. Robinson (1962, p. 60) notes that the propensity to accumulate in this phase of the business cycle may speed up the possible rate itself through induced innovations. An acceleration of the rate of accumulation with a prolonged expansion may be the result of a temporarily raised A curve. However, as the new equipment emerges, productive capacity rises relative to effective demand: 'The expectation of profit (projected from the current situation) deteriorates' (Robinson, 1962, p. 64). The development that until this new phase has been characterized by a higher level for the induced rate of profit than for the expected rate is now changed. The induced rate of profit has become lower than the expected rate of profit. This is the result of leftward change in the I curve.

This means that the propensity for capital accumulation has decreased. With a higher expected rate of profits it is now necessary to keep the rate of accumulation unchanged. The new situation of stagnation can also be strengthened by a rightward shift in the A curve. This means that the profit rate induced by a given rate of capital accumulation has been reduced. This may be the result of a reduction in the income multiples by a decrease in the propensity to consume.

Such changes in the propensity to consume and invest depend on the degree of capacity utilization when the rate of accumulation starts to go down. If capacity utilization is high, production costs will rise. It is more likely that reduction in the propensity to invest rather than reduction in the propensity to consume causes the cyclical change from expansion to stagnation. The stagnation phase can be prolonged by reduction in the propensity to consume, particularly if reduction in production growth is combined with rising costs — the so-called short-term Verdoorn effect. Robinson (1962, p. 58) analyses such a development and calls it 'a bastard golden age'. Inflation is combined with high unemployment if the real wage rate cannot be depressed sufficiently to obtain an acceptable cost growth. This development was characteristic of the industrial world after the two oil price crises during the 1970s and led to restrictive economic policy in most countries. The change from a fairly high growth rate in several European industrial countries to stagflation after the two oil price crises can be illustrated using Robinson's model.

The situation before 1974 was characterized by an increasing accumulation rate, so that induced rate of profit was higher than the expected profit rate. At the actual rate of accumulation a_a the A curve was above the I curve (Fig. 3.4). The restrictive policy moves the A curve to the right as the profit rate induced by the accumulation rate is reduced. At the same time profit expectations are strongly reduced, which means that the I curve moves to the left. The result is that the rate of accumulation at a_a will be on the right-hand side of the new equilibrium level and there will be a continuous decrease in the rate of accumulation towards this new equilibrium.

In the United States this development was opposed by a combination of a very restrictive monetary policy, which led to a strong international reduction in production and employment, and an expansionary budget policy with a high growth rate for public expenditure and tax reduction for the business sector in order to stimulate investments. The result was a declining rate of inflation and unemployment but increasing deficits in the balance of payments and the federal budget. It also led to such large deficits in the Latin American countries that financial collapse threatened Mexico in 1982 and the United States had to modify its economic policy.

Robinson's model can be used to analyse the US policy for fighting inflation and stagnation if the restrictive monetary policy which reduced

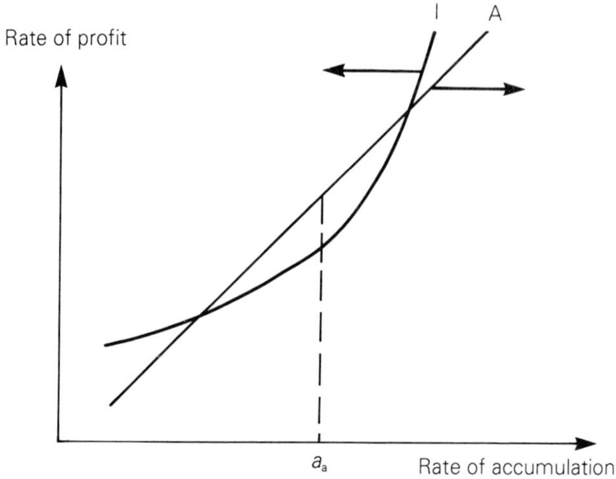

Fig. 3.4 Effect of cyclical variations on the relations between rate of profit and rate of accumulation.

both internal and international demand and thus increased the general propensity to save is represented by a reduction in the income multiplier and thus moves the A curve to the right. The US financial policy with higher public expenditure and lower taxation in the business sector can be interpreted as a movement of the I curve to the right. The displacement of the I curve is sufficient to eliminate the restrictive movement of the A curve to the right. The disequilibrium position of the US economy is characterized by a situation where the induced rate of accumulation is higher than the expected rate of accumulation, and the economy will have an increasing rate of accumulation and growth and a decreasing rate of unemployment.

3.4 THE STAR MODEL

STAR is a French model for economic planning which was strongly influenced by Robinson's theoretical work (Ministère de l'Economie et de Finances, 1974).

An abbreviated version of the model consists of 14 central relations:

$$p_{\mathrm{Y}} + \bar{p}_{\mathrm{m}}M = p_{\mathrm{I}} + p_{\mathrm{c}} + \bar{p}_{\mathrm{G}} + \bar{p}_{\mathrm{Ex}} \tag{3.1}$$

$$\frac{p_{\mathrm{I}}}{K} = a_1\frac{\mathrm{Aut}}{K} + a_2\frac{\mathrm{PAS}}{\mathrm{Aut}} + a_3 \tag{3.2}$$

$$p_{\mathrm{c}} = c_1 R + c_2 \bar{D}_{\mathrm{m}} + c_3 \tag{3.3}$$

$$M = m_1 Y \tag{3.4}$$

$$\frac{W}{p_y} = s_1 G\left(\frac{\widetilde{K}}{N}\right) + s_2 \text{ CONFLITS} + s_3 \tag{3.5}$$

$$N = g\left(y, \frac{\widetilde{K}}{N}\right) \tag{3.6}$$

$$\frac{\text{Aut}}{K} = b_1 \frac{Y}{\widetilde{K}} + b_2 \left(\frac{\text{PAS}}{\text{Aut}}\right)_{-1} + b_3 \tag{3.7}$$

$$\text{Aut} = p_y(1 - t_e) - W \tag{3.8}$$

$$R = (I - t_m) W \tag{3.9}$$

$$\text{pas} = f(\text{PAS}_{-1}, R - p_c + \bar{D}_m, \bar{p}_m M - \bar{p}_{Ex}, \text{tep } y + t_m W - \bar{p}_G) \tag{3.10}$$

$$\widetilde{K} = \widetilde{K}_{-1}(I - K) + I_{-1} \tag{3.11}$$

$$K = p_{-1}\widetilde{K} \tag{3.12}$$

$$\text{PAS} = \text{PAS}_{-1} + \text{pas} \tag{3.13}$$

$$p = \frac{p_y}{y} \tag{3.14}$$

The model includes 14 endogenous variables: nominal national product p_y; imports (volume) M; nominal investment p_I; nominal consumption p_c; gross saving of enterprises Aut; real capital K in nominal value; real capital $\widetilde{K}$ in constant prices; total liabilities PAS in year $t - 1$ of enterprises; net variation PAS in total liabilities of enterprises; national product Y in constant prices; price level p; wages W; employment N; household disposable income R. There are five exogenous variables: government expenditure p_G; exports p_{Ex}; net variation D_m in households' liabilities; number CONFLITS of labour market conflicts during the period; import price p_m.

The model is used to study how supply and demand determine the price level in the economy. The supply function is calculated by determining the business profit from two angles. First profits after taxes and other income transfers are calculated using the definition of factor incomes in Equation (3.8). According to Equations (3.5) and (3.6) wages are determined by employment and by the production volume. Therefore profits in the business sector after transfers to other sectors, i.e. the business sector's gross saving, are given by

$$\text{Aut} = (1 - t_e - s, (Y))\, p_Y \tag{3.15}$$

The business sector's funded consolidated profits can also be determined by the equation of the yield of capital. Equation (3.7) gives the business sector's profits after income transfers as a function of the national product in real terms and the financial position of the sector:

$$\text{Aut} = b_1' Y + b_2' \tag{3.16}$$

b_1' is the coefficient b_1 multiplied by the prices of the previous period:

$$b_i' = b_1 P_{-1}$$

b_2' is derived from the second factor in the function for the yield of capital (Equation (3.7)):

$$b_2' = \left[b_2\left(\frac{\text{PAS}}{\text{Aut}}\right)_{-1} + b_3\right]K$$

According to this function the yield of capital depends not only on Y/K but also on the relations between the passive posts of the business sector (debts and own capital) at the end of the previous year relative to the sector's gross savings during the current year. This relation increases the strength of the financial structure of business as an increase in the ratio indicates a need for higher debt financing.

Equation (3.15) gives the business sector's gross savings as a function of the nominal GNP and Equation (3.16) gives it as a function of the real GNP. Therefore the two equations together can give a relation between the price level and the real GNP. This relation is the supply function presented in the analysis of the STAR model by Mazier (1978, p. 218):

$$[1 - t_e - s_1 (Y)]\, p_Y = b_1' Y + b_2' \tag{3.17}$$

or

$$(A)p = A(Y) \tag{3.18}$$

The term (A) is positive at normal values for the coefficients, as expected for a supply function. Increasing prices are a condition for the business sector's willingness to augment production in order to defend its gross savings and its yield on capital. This can be said to be the conclusion of STAR regarding the supply behaviour of the business sector.

The total demand for finished goods can be calculated from the abbreviated presentation given in Equations (3.1)–(3.14) (Mazier, 1978, p. 215). We see from Equation (3.1) that the nominal value of public expenditure, the real value of exports and import prices are given exogenously. The nominal values of private investment and consumption are given by Equations (3.2) and (3.3) respectively. The real value of imports is given by Equation (3.4). The nominal value of household disposable income, which enters the consumption function, is given as a function of nominal wages in Equation (3.9). The nominal yield on capital, which enters the investment function, is given by Equation (3.7).

The investment demand is linearized as $p_I = a_1'\text{Aut} + a_2'$, where a_2' depends upon the credit possibilities (*'le retour financier'*) according to Equation (3.10). The business sector's gross savings Aut can be defined as the surplus, as was indicated earlier, and hence the investment function is written:

$$p_{I} = a_1'[1 - t_e - s_1(Y)]p_Y + a_2'$$

The consumption function is transformed using Equations (3.5) and (3.6) which give the wage share and employment as functions of the capital intensity and the national product. The wage share is also influenced by the number of conflicts in the labour market, which is an exogenous variable. The consumption function can be written

$$p_C = c_1(1 - t_m)s_1(Y)p_Y + c_2'$$

The transformed functions for investment and consumption demand are included in the equilibrium condition, which is given in Equation (3.1):

$$p_Y\{1 - a_1'[1 - t_e - s_1(Y)] - c_1(1 - t_m)\, s_1(Y)\} = c_2' + a_2 + \bar{P}_G x - \bar{p}_m m_1 Y$$

or

$$(B)p = B(Y)$$

This function, which represents the total demand for finished goods as a function of the price level, has a negative slope.

Finally we can study how different exogenous factors influence supply and demand and cause different types of inflation. Four types can be identified:

1. cost inflation–both curves are moved upwards;
2. profit inflation–the supply curve moves upwards;
3. demand inflation–the demand curve moves upwards;
4. credit inflation–the demand curve moves upwards.

With the traditional slopes of supply and demand curves there will be a stable equilibrium in each case.

An application of the STAR model to our analysis

It is a central theme of our production analysis that there are cyclical changes in demand and supply elasticities of different extents in different production sectors. In the industrial sector the supply curve is assumed to have a marked negative slope. A cyclical expansion will lead to a sharp increase in productivity growth. If prices are kept stable there will be an increase in profits. As the demand function will probably have a fairly low price elasticity and a high income elasticity in boom conditions, the result is an unstable equilibrium. Thus the expansion is strengthened.

In a stagnant economy with reductions in demand, the price elasticity is probably much higher than during a boom. The demand curve will be steeper. The possibilities for cost reduction will give a more elastic supply curve, and so the reduction in demand will move the economy to stable equilibrium characterized by lower employment and higher costs, which may lead to a stagflation with higher prices and unemployment. There is a risk that stagflation will be protracted. Both boom and recession will show peculiarities not present in STAR.

3.5 KALDOR'S ANALYSIS

Kaldor has emphasized the role of factor income distribution in economic analysis. He has followed its development from Ricardian theory through to Keynesian theory (Kaldor, 1956–57, pp. 83–100). His analysis of Keynesian theory is of particular interest for our study.

He emphasizes that Keynes regarded the profit share P/Y, the rate of profit in investment P/vY and the real wage rate $(W/P)/L$ as functions of the ratio I/Y of investment to production, and not the other way round (Kaldor, 1956–57, p. 97). Keynes treats the investment demand schedule ('the marginal efficiency of capital') as an independent variable (Kaldor, 1954, p. 54).

In order to analyse the relation between profit share and investment choice Kaldor does not use a model for cyclical variations but assumes equilibrium at a state of full employment. Income is divided between wages W and profits P. The model includes a saving function, in which saving is divided between capital owners' saving S_p and wage earners' saving S_w. The model is written

$$Y \equiv W + P$$

$$F \equiv S$$

$$S \equiv S_w + S_p$$

Investment is assumed to be given and the saving functions are assumed to be proportional:

$$S_w = s_w W$$

$$S_p = s_p P$$

The equilibrium condition gives

$$I = s_p P + s_w W = s_p P + s_w(Y - P) = (s_p - s_w)P + s_w Y$$

where

$$\frac{I}{Y} = s_p - s_w \frac{P}{Y} + s_w$$

and

$$\frac{P}{Y} = \frac{1}{s_p - s_w} \frac{I}{Y} \frac{s_w}{s_p - s_w}$$

The model operates only if

$$s_p \neq s_w$$

$$s_p > s_w$$

The latter condition is a stability condition (if $s_p < s_w$, a fall in prices would cause a fall in demand and thus generate a further fall in prices, just as a rise in prices leads to cumulative processes moving away from equilibrium). The degree of stability depends on the difference in the marginal propensities. The central idea of the model, i.e. that the Keynesian hypothesis makes the ratio of investment to output an independent variable and thus invariant with regard to the two propensities to save, the profit ratio P/Y, the rate of profit P/vY, and the real wage rate $(W/P)/L$, can only be true under certain conditions. Kaldor introduces the following restrictions.

1. The real wage cannot fall below a certain subsistence minimum. This condition is formulated as follows:

$$\frac{P}{Y} < \frac{Y - w'L}{Y}$$

   where w' indicates the subsistence minimum wage level. If the wage level falls below w', demand will be reduced and so there will not be full employment. Output will be limited by existing capital instead of labour. However, this development can be counterbalanced by technical developments which will lead to investment growth and a return to full employment.
2. The profit share cannot fall below what asset owners receive from alternative assets. The yield on real capital asset includes a risk premium and the rate of this yield is called the 'risk premium rate' r (Kaldor, 1956–57, p. 98). This condition can be written as follows:

$$\frac{P}{VY} \geqslant r$$

   There may also be other restrictions on the ratio of profit to capital because of the degree of monopoly in business. This minimum rate of profit is related to the business turnover and defines a minimum rate m for the profit ratio:

$$\frac{P}{Y} \geqslant m$$

If these two conditions for the level of the profit rate are unsatisfied, the assumption of full employment breaks down and the economy will stagnate.

There may be a number of causes for such a situation. Investment opportunities may be low because the full capacity growth planned by business is low relative to the 'natural growth rate' resulting from the growth in labour supply and technical development. In such a situation business expects a lower growth rate for future demand than that corresponding to potential demand determined by the growth of capacity. Another reason why the profit rate conditions may not be satisfied may be that the liquidity preference or the risks associated with investment are too high. Finally, lack of competition may cause 'over-saving' because of excessive profit margins. The result in this case will also be stagnation when there is an increase in the capital-to-output ratio. This may result in an increasing production growth rate.

Kaldor's analysis is related to Harrod's growth theory. His assumption of equilibrium is formulated in Harrod's terms: the warranted rate is assumed to be achieved. However, contrary to Harrod, he also assumes that at the warranted growth rate the economy has also reached full employment. Therefore the warranted growth rate G_w, which in Harrod's theory is defined as $G_w = s/v$ where s is the marginal propensity to save and v is the capital-to-output ratio, is also equal to the natural growth rate G_n. Kaldor states that the warranted growth rate and the natural growth rate are not independent of one another. If profit margins are flexible, the warranted growth rate will adjust itself to the latter through a subsequent change in P/Y.

Kaldor does not believe that the causes of cyclical movement are to be found in the lack of adjustment between s and Gv. In fact he does not agree with his own interpretation of the Keynesian model for explaining the business cycles. He states that the causes of cyclical movement should be sought in the disharmony between the growth rate G' desired by entrepreneurs, which governs the rate of increase of output capacity, and the natural growth rate G which depends on technical progress and the growth of the labour force. Figures 3.5 and 3.6 show that it is the excess of G' over G, and not that of s over Gv, that causes periodic breakdowns in the investment process because growth in output capacity outruns growth in production.

In another study Kaldor (1954) analysed the additions that have to be made to the Keynesian model in order to include the business cycle. In this analysis Kaldor does not consider the importance of factor income distribution in cyclical development. However, it is possible to include income distribution in his model. According to Kaldor the only change in the Keynesian model is to treat the investment demand schedule not, as Keynes did, as an independent variable, but as a function of both the existing stock of capital and the current rate of output (or simply a function of the ratio of current output to the stock of capital): 'This is justified, if future profit expectations can be regarded as being largely determined by the current rate

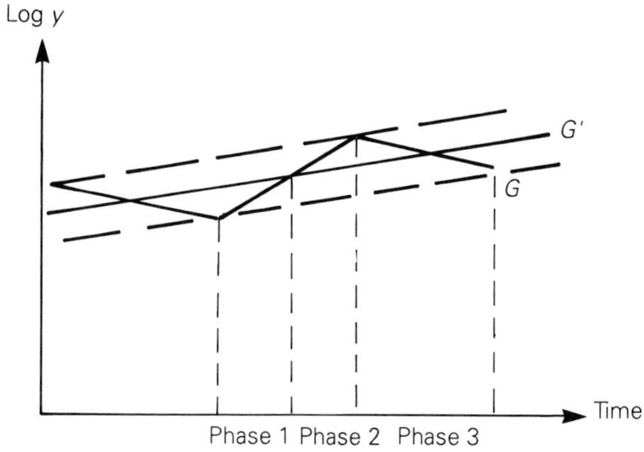

Fig. 3.5 Relation between natural growth rate and desired growth rate in Kaldor's theory. *G* is the natural growth rate (growth of labour force plus technical progress) and *G'* is the desired growth rate which governs the rate of output capacity. (From Kaldor, 1956–57, p. 97)

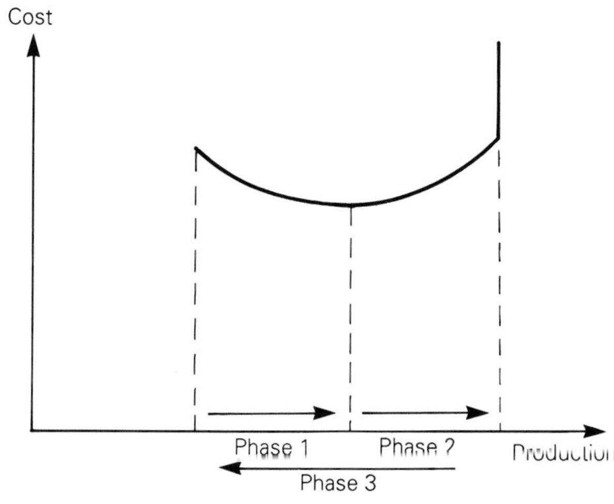

Fig. 3.6 Production costs in different cyclical phases.

of profit on capital' (Kaldor, 1954, p. 54). The current rate of profit per unit of capital (or rather per unit of investment) clearly depends on the current ratio of output to capital (at any rate if, as a first approximation, profit per unit of output can be taken as given) so that the rate of investment will increase with a rise in output.

This investment theory is combined with the traditional Keynesian multiplier theory according to which the increase in income created by a given addition to investment is greater the smaller is the marginal propensity to save. It is also assumed that there are restrictions on the rate of investment during the business cycle. In periods of stagnation gross investment can never be less than zero, and at the peak of the cycle the investment cannot be higher than what can be handled by investment production during full employment. Kaldor notes in this connection that the possibility of increasing investment by reducing consumption with the help of monetary inflation has been limited in practice as a reduction in consumption demand also tends to reduce the propensity to invest (Kaldor, 1954, p. 56).

Kaldor refers to a model for the trade cycle that he published in 1940 (Kaldor, 1940). In this model investment is a function of production, called the 'level of activity', and the stock of capital. He emphasizes that the dependence of investment on the production level should not be regarded as related to the acceleration principle. He assumes that changes in the current level of production influence the current level of profits which in turn influences investment demand (Kaldor, 1940, p. 79). This interpretation of what determines investment demand explains Kaldor's hypothesis that the relation between investment and the national product should not be linear over the cycle. It is probable that dI/dY will be small for low levels of the GNP relative to its 'normal' level because there is a good deal of surplus capacity. A rise in production and profits will not result in an increase in investment in real capital as long as there is still surplus capacity that can be used. The business sector will instead increase its investments in financial assets with a different degree of liquidity.

When the income expansion has reached the normal long-term growth level there will be a marked increase in investment demand as there is a need for new production capacity. The investment demand is stimulated by the increased growth in financial assets during the early part of the expansion. Profit expansion is also stimulated by the high stable growth rate for current production. When the income expansion reaches the top of the business cycle, increasing costs and increasing difficulties in borrowing will result in lower levels for dI/dY for the investment function. Kaldor (1954, p. 56) notes that the assumption of non-linear coefficients makes it unnecessary to rely on time lags or to introduce exogenous shocks in order to account for cyclical movements.

Figure 3.7 shows graphically the investment function in Kaldor's trade cycle model. The model is completed by the introduction of the savings function which is also characterized by non-linearity over the cycle. The marginal propensity to save is likely to be relatively large compared with its normal value for both low and high income levels. When incomes are unusually low, savings are reduced and below a certain level of income they become negative. When incomes are unusually high, savings will contain a higher proportion of income. This tendency will be stimulated as prices tend to rise faster than wages, and so there will be a shift in the distribution of

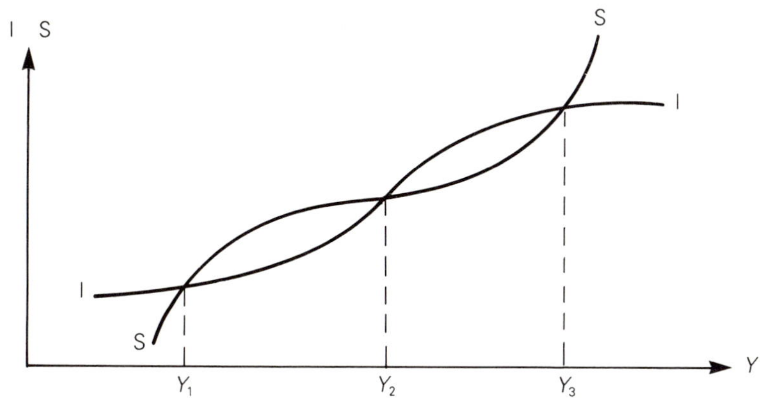

Fig. 3.7 Investment and savings functions in Kaldor's trade cycle model.

income in favour of profits and thus an increase in the aggregate propensity to save.

As Fig. 3.7 shows, Kaldor's trade cycle model has three equilibria for the GNP at Y_1, Y_2 and Y_3. The dynamic character of the model is not confined to the fact that the propensities to invest and to save depend on the development of the GNP over the trade cycle. The two functions include another variable, the real capital, whose development is dependent on the income growth rate and whose changes lead to movements in the investment and savings functions. At the low equilibrium level Y_1 the low level of gross investment is not sufficient to cover the depreciation in capital stock. Its value will go down, which will result in an upwards shift of the investment function. The savings function will tend to shift downwards because of the difficulties that households will experience in covering consumption needs at the very low level of income. These movements of the investment and savings functions will change the character of the equilibrium at Y_1. In its original form it is stable. However, after the two curves have moved it tends to be unstable upwards as the curves are just tangent at Y_1 and the I function tends to be higher than the S function above the tangential point as shown in Fig. 3.8. When the economy is on the right-hand side of the tangential point there will be expansion because of the disequilibrium, as curve I is above S until a new equilibrium is reached at the top of the trade cycle at GNP level Y_3. However, the change in capital stock at this level will also cause shifts in the investment and the savings functions. Investment leads to an increase in the capital stock and rising costs, and in this cyclical situation a restrictive financial development with increasing interest rates can result in a decreasing propensity to invest. Therefore the I curve will tend to shift downwards while the S curve may shift upwards as a result of the increase in capital formation. This can be

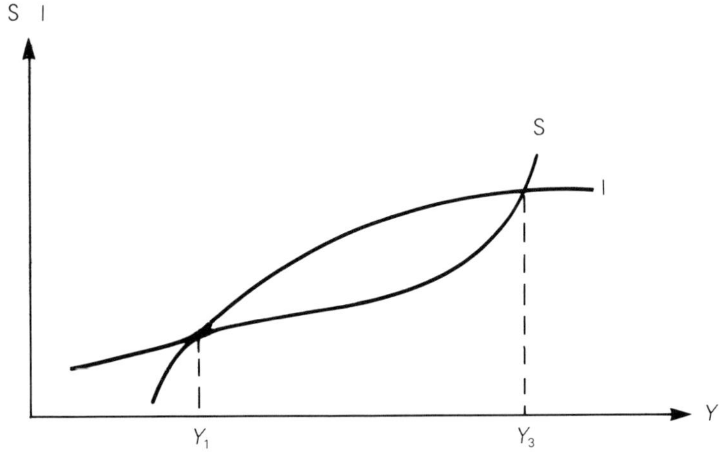

Fig. 3.8 Investment and savings functions during an expansion phase.

explained by a decrease in share prices as expectations of continuous income growth no longer hold. A decrease in household wealth may lead to a reduction in consumption expenditure and thus to a higher propensity to save. The shifts in the I and the S curves at the high income level will change the stability condition of the equilibrium at Y_3. It will no longer be stable downwards as shown in Fig. 3.9 and cyclical downturn will start. This will not stop until the economy has reached a new stable equilibrium at the low income level Y_1.

Kaldor notes that the important factor for the cyclical development is the capacity of the capital goods industries.

> The greater is this factor, the shorter the duration of the boom phase (both absolutely and relatively to the depression phase) and the greater the amplitude of the cyclical movement. The capacity of the capital-

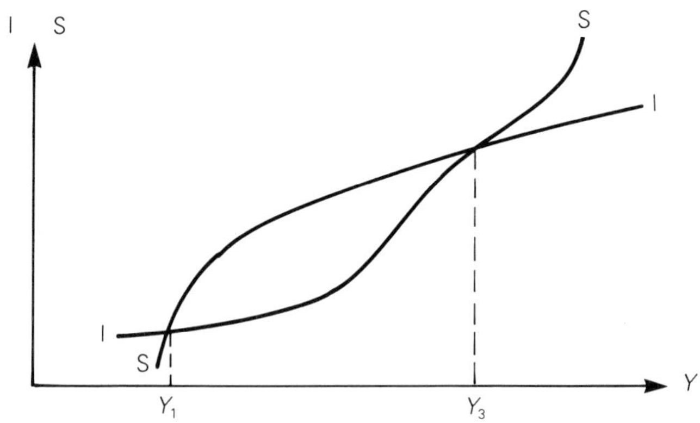

Fig. 3.9 Investment and savings functions during a contraction phase.

> goods industries relative to the consumer-goods industries may in turn be assumed to depend on the proportion of income saved at full-employment income levels—in other words, on the short run savings function. (Kaldor, 1954, p. 60)

This relation may have been important in explaining the change in the medium-term growth rate in some industrial countries in the latter part of the 1970s. The high growth rate during the 1960s and the first part of the 1970s resulted in a high growth rate for business income and saving. The good profit expectation led to high investment and a high capacity in the capital goods industries relative to the consumer goods industries. When a restrictive policy led to decreasing demand in the latter part of the 1970s this high capacity in the capital goods industry was believed to be too large and the result was a drastic reduction in the propensity to invest, a reduction which in some countries was not restored until the middle of the 1980s.

Kaldor has only hinted at the importance of the changes in factor income distribution for explaining this dynamic process. Such an analysis could be included in the model by comparing the development of the growth rate desired by entrepreneurs and determined by household investment and saving behaviour with the natural growth rate depending on labour output and technical development.

The graphical presentation in Fig. 3.10 combines Kaldor's trade cycle model with a diagram that shows the development over the trade cycle of the natural growth rate G_n and the growth rate G' desired by entrepreneurs and households which governs the increase in output capacity. The figure shows that the low-income equilibrium level in the trade cycle model is reached at time t_0 and the cyclical peak is reached at time t_1. This is followed by a period of stagnation with a lower growth rate, which lasts until time t_2 when a new boom starts.

The cyclical development of G_n can be compared with the more stable growth of G'. The figure has been simplified to show a constant growth rate for G'. In fact G' is influenced by the development of G_n, but the important point is that G' has much smaller movements compared with those of G_n. During the whole expansion period from t_0 to t_1 the natural growth rate G_n is higher than the entrepreneurs' desired growth rate G', which governs the rate of increase of output capacity. In industrialized countries G_n is generally a result of a high productivity growth rate. The development between t_0 and t_1 is therefore characterized by an increasing profit share for most of the time. The result will be an increasing propensity to invest which, as can be seen from Fig. 3.10(a), accelerates when the capacity utilization has increased sufficiently. In Fig. 3.10(b) such a situation is assumed to have been reached at t'_1.

Before reaching the maximum boom level Y_1 at t_1, the profit share peaks

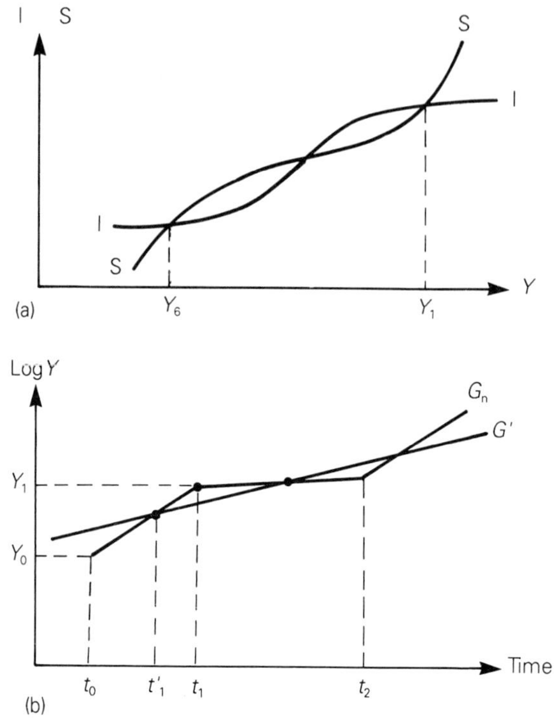

Fig. 3.10 Kaldor's trade cycle model combined with the relation between natural and desired growth.

and begins to decline. It reaches its lowest level during the period of stagnation between t_1 and t_2. The increase in the wage share of the GNP is due to the decrease in productivity which in turn is due to the reluctance of firms to decrease their labour force when production is decreasing. This is an effect of the high cost of recruiting and training labour when the production starts increasing again. This is called the high transaction cost for the labour force and has been much emphasized in modern literature (see, for example, Faxén *et al.*, 1989).

The development of factor income distribution in Kaldor's business cycle model can be illustrated by Fig. 3.10(b) which shows the development of G_n and G' during the cycle. The growth rate G_n influences producers' expectations of future growth rates and is therefore the basis for their long-run plans for future growth of production capacity symbolized by G'. G' also represents the long-run growth rate for producers' labour costs. During expansion G_n is greater than G'. This means that the producers' profit share is growing. When the economy has passed the cyclical peak Gn is less than G'. This means that the profit share is decreasing. If the cyclical expansions and recessions are of about the same length and scale, the cyclical changes in income distributions will cancel out and the medium-term income distribution will be stable if there are no structural changes due to other factors. Such changes have occurred in several industrial countries and are the result of medium- or long-term changes in the production structure.

(a) *Growth in Kaldor's trade cycle theory*

Kaldor believes strongly that it is not necessary to assume economic growth in order to explain cyclical fluctuations. Rather, economic growth should be treated as a cause of aberrations from the pure rhythm of the static cycle than as a *sine qua non* without which the basic phenomena could not be explained. Long-run growth is explained by the increased capital creation in successive booms that has exceeded the net capital depletion in periods of depression.

The central factor influencing long-run development is how entrepreneurs and households interpret the actual short-run development and how this development influences their long-run expectations, for this will determine the desired growth rate which governs the rate of increase in output capacity. This is the growth rate denoted G' in Fig. 3.10. A high value for G' leads to increasing capacity in investment goods industries, which in turn secures a high capacity in comsumption goods industries in later booms.

Kaldor considers that the continuous growth of production during successive booms is the result of population growth or technical progress. Technical progress can be neutral in the sense that the ratio of capital to output remains constant in the medium term or even in the long run. However, it should be noted that such a development is only a historical accident and should not be considered as a relation that explains long-run growth. He concludes that it is not the trend rate of growth that determined 'the strength or duration of booms, it is the strength and duration of booms which shapes the trend rate of growth' (Kaldor, 1954, p. 68).

The relation between the current short-run growth rate of expenditure, the expected future growth of expenditure and the rate of increase of output capacity is illustrated graphically in Fig. 3.11. The linear natural growth rate during the past (from $t-1$ to t_0) results in expectations of a continued high natural growth rate G_n^e in the future. These expectations lead to an increased growth rate G' for planned output capacity. The expected high growth rate will lead to an expectation of high profit growth and stimulate an increase in investment. During the boom there will be an increasing profit share, which will lead to an increase in the capital intensity. Thus there will be a long-run Verdoorn effect, which has been analysed earlier in this chapter.

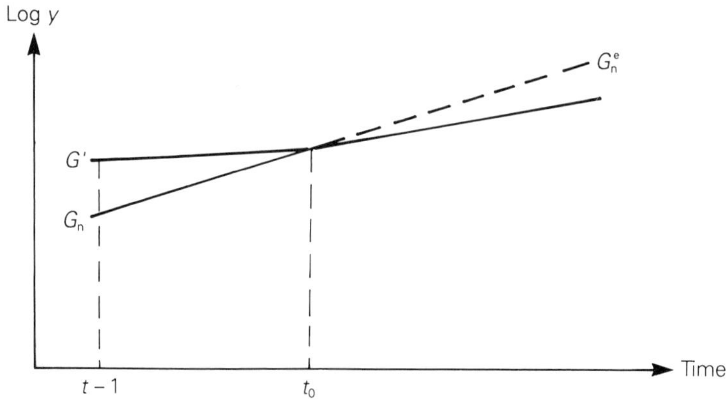

Fig. 3.11 Expectations when natural and desired growth differ.

According to Kaldor the causes of the changes in factor income distribution in the medium and long term should be sought among the factors which change the extent of the business cycles. These are primarily related to changes in the cyclical pattern in the development of the capacity utilization.

(b) *The neoclassical viewpoint*

As a contrast to the preceding discussion we can consider the neoclassical view of the development of the factor income distribution in the long run. Short-run variations are not considered in this theory as a difference between the actual growth rate and the full capacity growth rate is not possible because of the assumption of perfect competition. It follows from this assumption that there cannot be any discrepancies between planned saving and investment. The production function and the savings function together with the exogenous natural growth rate are sufficient to determine the factor income distribution in the long run.

The neoclassical approach can be illustrated using an example in which effective man-hours E_t grow at a rate $g_L + \lambda$, where g_L is the rate of growth of the labour force, and λ is the rate of growth of the effective labour units per man-hour. This gives

$$E_t = L_t \exp(\lambda t)$$
$$= L_0 \exp[(gL + \lambda)t]$$

The production function in terms of real capital and the effective labour force can be written

$$Q_t = F(K_t, E_t)$$

The production function exhibits constant returns to scale, and so we can divide by E_t to obtain the labour-intensive production function:

$$\frac{Q_t}{E_t} = f\left(\frac{K_t}{E_t}\right)$$

A savings function is included, in which savings is a constant fraction s of output. Thus

$$\frac{\mathrm{d}K}{\mathrm{d}t} \equiv I_t = S_t = sQ_t$$

The growth rate of the capital intensity equals the difference between the growth rate of the capital stock and the growth rate of the effective labour force:

$$\frac{\dot{K}}{E} = \frac{sf(K/E)}{K/E} - (g_L + \lambda)$$

There will be a stable equilibrium value $(K/E)^*$ for (K/E), which is obtained when $\dot{K}/E = 0$, so that

$$f\left(\frac{K}{E}\right)^* = \frac{g_L + \lambda}{s}\left(\frac{K}{E}\right)^*$$

A graphical representation of this equilibrium is shown in Fig. 3.12(a). Therefore in equilibrium with a constant value for Q/E and K/E, output per man-hour and capital per man-hour will grow at the rate of the technical growth factor λ. The relative shares of factor income will be constant. This can be shown in Fig. 3.12(a) by introducing a tangent whose slope indicates the marginal product of capital p.

Therefore in neoclassical growth theory the factor income distribution is given provided that the production functions, the savings function and the natural growth rate are given. Is it now possible to analyse the result of change of income distribution?

A change in the factor income distribution can be assumed by introducing a break in a long growth process which has been characterized earlier by a constant technical growth. This technical growth is assumed to be Harrod neutral, so that the capital-to-output ratio is constant. However, at times this process is interrupted when bottlenecks in the production process result in lower production growth. The result is a change in factor income distribution as the share of profits decreases. This development is illustrated in Fig. 3.12(b) which shows how stable technical growth moves the F function from F_0 to F_1. However, instead of continuing to F_2, the growth rate is reduced so that the function is located at F_2' instead. It can be seen that the F_2' curve is less steep at the given long-run capital output ratio V_0, which indicates a decrease in the profit share.

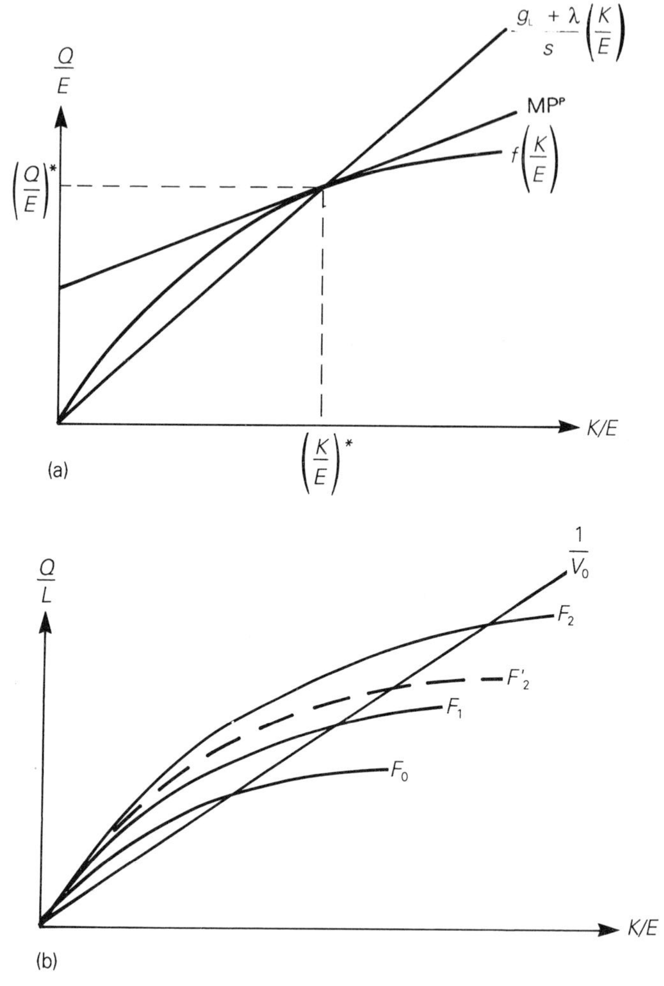

Fig. 3.12 (a) Factor income distribution in a neoclassical equilibrium; (b) change in factor income distribution as a result of technical change.

However, it is difficult to interpret this sudden break in the growth process and its consequences for factor income distribution as an unexpected event that will have serious effects on the financial situation and on the propensity to invest in the business sector, for the rules of the game in neoclassical theory exclude changes in capacity utilization and income expectation. The economy is as much in equilibrium at the new lower growth rate as it was at the earlier one. We shall therefore have to seek other approaches to the analysis of long-run changes in factor income distribution.

3.6 STRUCTURAL CHANGE AND LONG-RUN DEVELOPMENT IN FACTOR INCOME DISTRIBUTION

Earlier in this chapter we considered how changes in the production structure influence the development of productivity and costs which leads to shifts in the aggregate supply curve. This structural change is particularly concerned with the increasing share of services in the total production in industrial countries. This development appears to lead to decreasing growth in production and productivity, which in turn results in lower profit shares.

As a conclusion these tendencies towards lower profit shares because of the changes in production structure in the industrial countries can be compared with the stimulating effects of capital investment and capital intensity through its effects on technical growth and productivity. The two counteracting factors can be illustrated graphically. Figure 3.13 shows how the two main factors influencing the profit share in the medium and the long term compete with each other. The increase in capital intensity which characterizes the latter parts of boom periods in particular leads to a higher productivity growth rate. However, this cyclical expansion also stimulates structural change in the economy. It leads to expansion in both the public and private service sectors. In the latter, services which are directed towards the industrial sector in particular tend to expand. However, this structural change results in lower average productivity growth rate and thus to a lower profit share.

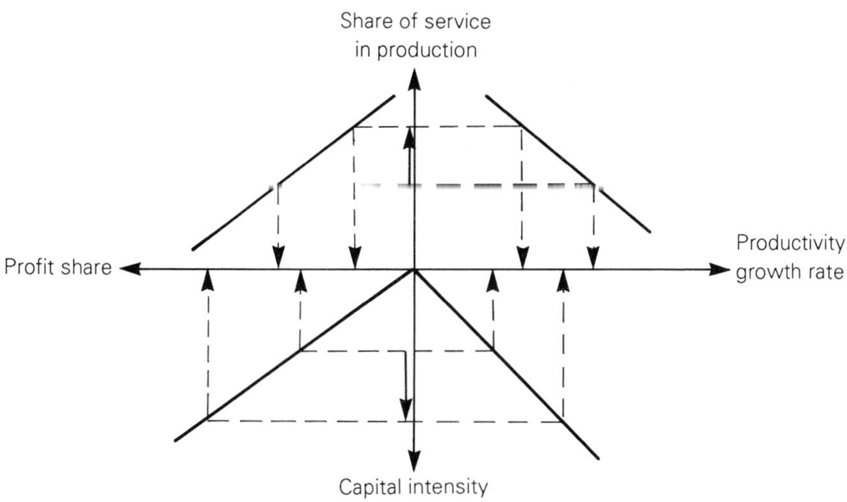

Fig. 3.13 Effects of changes in production sector composition on factor income distribution.

REFERENCES

Boltho, A. (1982) Growth, in *The European Economy* (ed. A. Boltho), Oxford University Press, Oxford.

Evans, M. K. (1969) *Macroeconomic Activity*, Harper and Row, London.

Faxén, K. O., Odhner, E. E. and Spånt, R. (1989) *Lönebildningen i 90-talets Samhällsekonomi*, (*The Wage Formation in the Economy of the 1990s*), Rabén and Sjögren, Simrishamn.

Kaldor, N. (1940) A model for the trade cycle. *Economic Journal*, March.

Kaldor, N. (1954) The relation of economic growth and cyclical fluctuation. *Economic Journal*, **64**.

Kaldor, N. (1956–57) Alternative theories of distribution. *Review of Economic Studies*, **23/24**.

Kalecki, M. (1965) *The Theory of Economic Dynamics*, George Allen and Unwin, London.

Ministère de l'Economie et de Finances (1974) Le modèle star, statistiques et études financières.

Lundberg, E. (1968) *Instability and Economic Growth*, Yale University Press, New Haven.

Mazier, J. (1978) *La macroéconomie appliquée*, Presses Universitaire de France, Paris.

Mazier, J. (1982) Growth and crisis—a Marxist interpretation, in *The European Economy* (ed. A. Boltho), Oxford University Press, Oxford, pp. 151–61.

Mitchell, W. C. (1950) *Business Cycles and their Causes,* University of California Press, Berkeley and Los Angeles, CA.

OECD (1989) *Economic Outlook, Historical Statistics 1960–87,* OECD, Paris.

Robinson, J. (1962) *Essays in the Theory of Economic Growth,* Macmillan, London.

Schumpeter, J. A. (1939) *Business Cycles*, McGraw Hill, New York and London.

4

Sector expenditure on final goods: short-run and long-run changes in the demand structure

4.1 CONSUMPTION DEMAND

(a) *Household consumption expenditure*

The current expenditure on final goods for period t is registered in the B matrix of the transactions table on page 13. The major part of this expenditure is private consumption. We find this expenditure in the household sector column and in the rows of those production sectors who have supplied the consumption goods and services to the household sector.

The household sector's rows in the matrices for factor incomes (the Y matrix) and for financial transactions (the F matrix) show the total current liquid resources which the household sector has at its disposal. They consist of the sector's current wages and the income from household enterprises together with the transfer income that the sector has received in the form of interest and dividends, payments from insurances, government transfers and subventions etc. They also include credits that the household sector has received from banks and other credit institutes, from the goverment sector etc. The household sector has used these current liquid resources for those financial and real expenditures listed in the household sector column in the matrices for financial transactions (the F matrix) and final goods (the B matrix). As already mentioned the expenditure for private consumption is recorded in the B matrix. Transfers and credits to other sectors, such as taxes to the government sector, insurance, pensions and interest payments to credit institutes, purchases of bonds and equities from government and business sectors, and changes in deposits in banks, are recorded in the household sector column in the F matrix. In summary, the household sector's use of its current liquid resources can be divided between transfers, saving and consumption expenditures.

(b) *Consumption development and consumption theory since the Second World War*

The economic expansion that had been the result of 4 years of war in both the western and the eastern hemisphere ended the problem of mass unemployment that had been so acute for a long time between the two world wars. The major problem was whether full employment could be maintained when peace returned. At the end of the Second World War a macroeconomic theory for the determination of total demand had been developed which had not been available at the end of the First World War. This theory could now be used to forecast post-war demand.

Expenditure played a central role in Keynes macroeconomic theory for household consumption. Keynes noted that the household is disposed to increase consumption as real income increases, but not by as much as the income increase. This means that the marginal propensity to consume is positive and less than unity. This theory has been of great importance in explaining the role of household expenditure in the development of income and production in different phases of the business cycle, for it means that the household consumption expenditure tends to smooth out the more violent changes in production and income produced by changes in expenditure on other types of final goods such as investment or export. In order to forecast how demand would develop after the war, it was the future stability of the propensity to consume which was of highest importance. In the United States an estimation was made of what private and government expenditure had to be in order to ensure full employment (Smithies, 1945).

Consumer expenditure was calculated on the basis of data from 1923 to 1940 for the relation between per capita consumption and per capita income, both deflated by the cost of living index. These estimations confirm the Keynesian hypothesis about the relation between consumer expenditure and household disposable income (Smithies, 1945, pp. 5–6).

As well as estimations of the post-war government expenditures it would in principle be possible to calculate the necessary increase in gross capital formation in order to reach a level of the total demand that could balance the estimated GNP production capacity at full employment. Of course, the authors were well aware of the very rough calculations of both the main types of demand for finished products as well as the estimates of the labour supply and productivity growth, which are the basis for the calculations of full capacity GNP. Anyhow the very large increase in private capital formation calculated to be necessary to bridge the gap between total demand at full capacity output convinced them that a very expansionary policy was necessary after the War if the United States was not to return to the same type of mass unemployment that had characterized most of the 1930s.

We know that development in the United States after the Second World War was quite different from these forecasts. There was full employment and, at least up to 1948–9, demand tended to be too strong instead of too weak. Why were the calculations of future demand wrong?

It had been shown that the main weakness in these forecasts was the way in which Keynesian consumption theory had been used (Dusenberry, 1952, pp. 70–6). Dusenberry (1952, p. 69) notes that '... regressions based on short-period data have been given a long period interpretation. It has generally been assumed that a regression between saving and income indicates what the long period consumption function would be in the absence of trend factors'. He also notes that by using regression between savings and income from the data for the period 1923–40 and applying this relation to an expected real income increase after the war of about 40%, they obtained estimates which were much too high and thus their forecasts for consumption development were much too low. His conclusion regarding the reason for this mistake is that cyclical fluctuations in income have an effect on saving which is entirely different in character from the effect of secular increases in income (Dusenberry, 1952, p. 70).

The contracyclical stabilization policy from the end of the Second World War up to the 1960s in the Western industrial countries was particularly concerned with the tendencies towards too rapid a growth in demand which resulted in cyclical changes in foreign trade and deficits in the balance of payments. During the first 10 years of this period the most important instruments of this policy in the industrial countries of Western Europe were regulation of imports, wages, prices and dividends. The interest rate was kept fixed at a very low level. This was partly due to the policy adopted for financing the budget deficit since the War and partly a weapon against an expected depression. This was planned to be counteracted by house-building projects in order to increase demand and keep down unemployment. In the United Kingdom the large budget deficit during the War had been financed without increasing interest rates. This had been possible as the banks had to redeposit their surplus of investable funds with the Treasury, thereby limiting the amount of credit that could be extended to the private sector. After the middle of the 1950s there was a tendency to replace regulations by monetary and fiscal measures. In 1955 there was a fiscal relaxation combined with a monetary tightening in the United Kingdom and other European countries. Although in this period monetary policy played a different role from that proposed by Keynes during the inter-war period, the active fiscal policy adapted to the cyclical situation conformed to Keynesian theory. It was an application of Keynesian multiplier theory and thus of short-term consumption theory.

A refined form of Keynesian short-term consumption theory can be found in Kaldor's model for the trade cycle (Kaldor, 1940, p. 78). The point of departure is Keynesian saving theory, in which saving is a positive function of the level of activity. However, this relation is linear only at an intermediate stage of economic activity. The marginal propensity to save is dependent on the level of activity. Kaldor notes that the marginal propensity to save 'is likely to be relatively *large*, both for low and for high levels of activity, as compared with its normal level' (Kaldor, 1940, p. 81). Such changes in the marginal propensity to save can be explained by household consumption behaviour. In a recession households try to maintain habitual consumption patterns. This may be done by drawing on savings from earlier periods. It may also be financed by unemployment benefits. When the income growth is high and the business cycle is at its peak the propensity to save will be large, because there will be a shift in the distribution of income in favour of profits, and so an important part of the increase in saving will be business saving. Kaldor illustrates the deviation of the savings function graphically (Fig. 4.1).

The function shows the short-run relation between savings and the level of activity Y, as Kaldor denotes the real national product. In a longer time perspective the saving function will not be stable. When the activity is high, a rise in production capacity will also lead to higher savings as well as higher consumption capacity. The curve will tend to move upwards. At low income levels there will instead be a gradual reduction in capital which will result in even lower real income and thus in lower savings. The savings function will move downwards. We shall return to the development of the propensity to consume and to save in the medium and the long term at a later stage of our analysis.

In the 1960s economic development in the industrial countries seemed to take a new direction. The cyclical movements seemed to be less marked and the short-term stabilization policy was able to keep unemployment down when activity decreased and also to stabilize price increases when economic growth accelerated. However, in the latter situation there were difficulties in

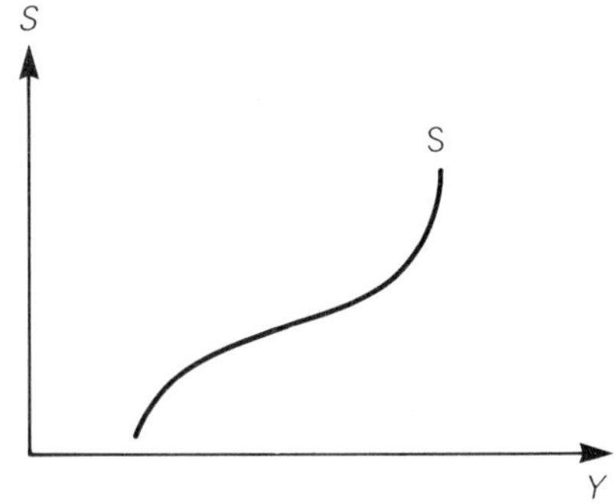

Fig. 4.1 Kaldor's saving function.

many countries because of deficits in the balance of payments. In general, there was a tendency to pay more attention to problems concerning medium-term and long-term development. In such a perspective the growth rate of production capacity and of capital formation and their effects on technical change became particularly important. The consumption theory was still central, but interest shifted to its long-run formulation.

The basis of increasing growth in the industrial countries was the need for rebuilding the war-devastated economies. This expansion continued during the 1960s mainly because of the increase in international trade when import regulations were removed and new organizations for economic co-operation and free trade, the European Economic Community (EEC) and the European Free Trade Association (EFTA), were created in Western Europe. A prerequisite for this development was the system of fixed exchange rates between the currencies in the leading industrial countries which had been created at the end of the War.

The economic expansion led to migration all over the industrial world. In Europe people migrated from the east and the south to the traditional industrial regions in the FRG, France, the United Kingdom, Belgium and the Netherlands. However, new regions were also industrialized. This development meant not only a rising real income growth in households, but also important structural changes in the distribution of consumption goods and services.

Even when the growth rate of private consumption in the industrial countries had been lower than the growth of GNP, so that the share of private consumption decreased, the slow growth rate of the population made possible a long-run increase in per capita consumption. The most typical change in the international pattern of consumption has been a higher share of durable goods relative to non-durable necessities. The increase in high income goods includes both goods and services, motor cars, sports wear, international travelling and tourism. If the medium- and long-term development of consumption shows such an important structural change, a strong complementary tendency between goods and services with a high income elasticity can be said to exist during periods of high real income growth.

Consumers in the household sector have been said to have a high habit dependence (Dahlman and Klevmarken, 1971, p. 149). The development of new housing areas outside urban centres creates the need for motor car transportation and durable household equipment for easy and cheap provision of services which during an earlier period had been handled by servants. Shorter working hours allow time for more sports and recreation.

The development of private consumption in Sweden during a period of high growth rate is illustrated in Fig. 4.2 which shows the development, in 1980 prices, of total private consumption and six types of goods and services (motor cars, other durable goods, housing, other services, food and

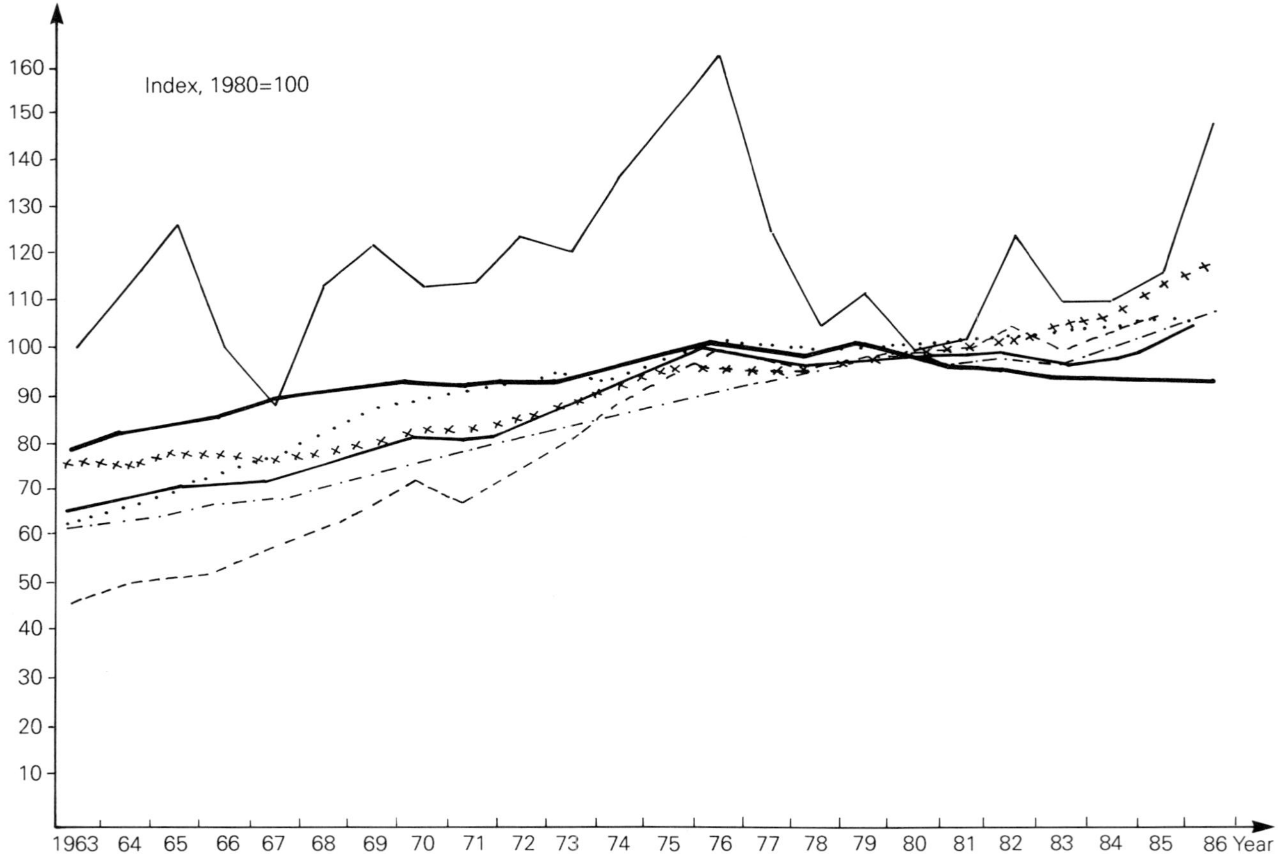

Fig. 4.2 Consumption development in Sweden between 1963 and 1986

other non-durables). The years 1963–76 represent a period of fairly high growth rate. The figure shows the high growth rate for durable goods other than motor cars, housing and other services during this period. The only group with a low growth rate was food. Motor cars showed a very unstable development, with periods of very high expansion interrupted by periods of marked decline. In general, however, private consumption showed a uniform expansion in goods and services demanded in a high-income industrialized society.

These tendencies to complementary relations between consumption goods and services with high income elasticity have been analysed using the linear expenditure system (Dahlman and Klevmarken, 1971, p. 34). The linear indifference curves in Fig. 4.3 indicate a stable relationship between the consumption C_1 and C_2 of two goods. The proportionality between the consumption of the two goods leads to a linear shift to new indifference curves as the real income increases from Y_1 to Y_2, as shown in Fig. 4.3.

When the growth rate went down in many industrial countries in the middle of the 1970s, there was also a stagnation in consumption growth (OECD, 1988). A new tendency was observed in the relation between consumption development for different types of goods and services. In Sweden stagnation mainly occurred in non-durable goods and services, but the consumption of durable goods steadily increased in the long run. The tendency to maintain consumption growth for housing and durable goods despite a low real disposable income growth has also been apparent during much of the 1980s.

Figure 4.2 shows how a slow growth rate for total private consumption in Sweden from 1976 to 1984 was followed by a period of higher growth from 1984 to 1986. The only group that showed a practically constant growth rate compared with the period from 1963 to 1976 was housing. For the other groups there was a period of stagnation from 1976 to the beginning of the 1980s. After 1984 motor cars and other durable goods, in particular, were characterized by an increasing rate. The increasing growth in consumption expenditure on durable goods during the 1980s in Sweden

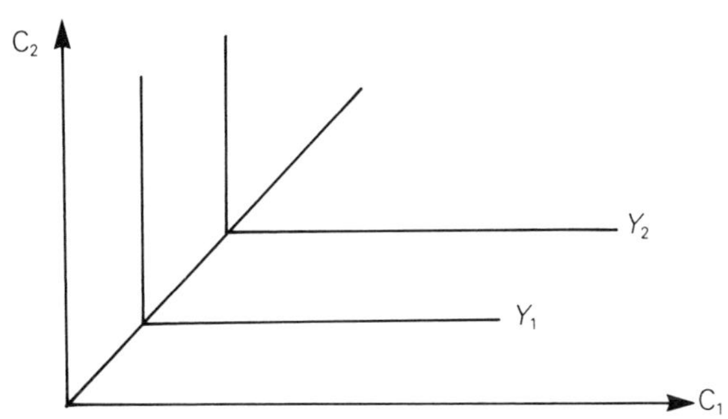

Fig. 4.3 Fixed consumption technology.

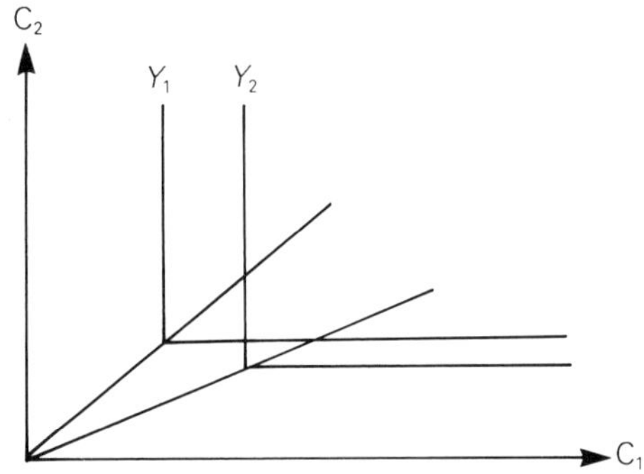

Fig. 4.4 Changes in consumption technology.

is surprising because of the stagnation in households disposable income during the period 1976–86. However, the good liquidity position of the banks made it easy for households to obtain bank loans and the possibility of obtaining a tax reduction by deducting debts in the declaration income of made credit financing of durable consumption expenditure extremely popular.

The new tendency in consumption demand that characterizes the period from 1976 to 1986 can be analysed using a linear expenditure system. Figure 4.4 shows how, despite a very low increase in income, households have been able to maintain the growth rate of housing and other durable consumption expenditures by using bank credits. Households have instead decreased their consumption expenditure on food, and this has been characterized by particularly high price increases. In Fig. 4.4 C_1 stands for housing and other durable goods and C_2 stands for food. The figure shows how the stagnation in real income growth has led to substitution. In order to maintain necessary growth in expenditure for housing and other durable goods, which has been possible with the help of bank credits, households have had to slow down their food consumption, for instance by changing to cheaper and less refined food products with more home preparation etc.

(c) *Regression analysis for consumption expenditure in Sweden*

A consumption function estimated on time series running from the first quarter of 1961 to the fourth quarter of 1986 was published in 1988 (Ekstedt and Westberg, 1988a). The importance of disposable income and unemployment gives the function a Keynesian character. In fact, the positive relation for unemployment and the negative relation for the change in unemployment makes it possible to interpret it in accordance with Kaldor's hypothesis. It agrees with his idea of a non-linear relation between consumption and real income, as stagnation in income growth with increasing unemployment tends to increase the propensity to consume. High income growth will lead to a lower propensity to consume as a state of saturation will arise.

The market value of the bond stock has particular importance as it shows how consumption is influenced by cyclical variation in bond prices. The relation is negative as time lags in household income postpone the expansion in consumption to a phase in the cycle when bond and share prices have generally declined. It should be added that the parameters show a high degree of robustness which implies that there is no major mis-specification.

In a study presented by Ekstedt and Westberg (1988b) the consumption function has three variables: change in income, level of disposable income and wealth relative to debt in the household sector. There is a negative relation between consumption and real income change. This shows the stiffness of consumption behaviour in accordance with Keynesian theory. There is a practically positive proportionality between consumption and household disposable income. The importance of the ratio of wealth to debt in households introduces an important argument from the life cycle theory. However, it should be remembered that the importance of wealth was also pointed out in Keynesian theory, particularly in the form presented by Kaldor, where long-term shifts in the consumption function were provoked by changes in wealth. The elasticities and standard errors of the parameters of the three variables in the consumption function are shown in Table 4.1.

Table 4.1

Variable	*Parameter*	*Estimate*	*Standard error*
Income change (D log Y)	β_1	−0.732	0.239
Disposable income ($Y - T$)	β_2	1.012	0.048
Wealth/debt (H_W/H_D)	β_3	0.136	0.035
Period 1963–76			
F value 599.1, adjusted R^2 0.99, dep. mean −80.69			
Disposable income		107.15	7.52
House prices		−12.56	7.82
Interest on bonds		0.08	2.61
Share prices		3.10	2.02
Intercept		1.82	
Period 1977–86			
F value −6.4, adjusted R^2 −0.70, dep. mean 100.58			
Disposable income		53.24	26.35
House prices		2.22	7.54
Interest on bonds		−9.09	7.78
Share prices		−1.12	1.89
Intercept		1.82	7.17

It has also been possible to illustrate the change in consumption when economic growth stagnated in the middle of the 1970s. An analysis of the relations between total consumption expenditure and the four variables household disposable income, interest rate, equity prices and house prices for the period 1963–76 on a yearly basis indicates the dominant role of disposable income in a period of high growth. There is also a positive relation between equity prices and consumption, which is expected in a period of high growth and rapid expansion in house building. It is also characteristic that in a period of high liquidity, the importance of the interest rate is hardly noticeable. Table 4.1 shows the good statistical results of the estimation.

A study of the same type for the period 1977–86 gives very poor results. It is characteristic that when the growth rate is low the parameter estimate for disposable income becomes much lower. The most interesting result of this estimation is the strong negative relation of the interest rate. It indicates the importance of the availability of bank credit in the development of consumption. In a situation of easy credit, increasing house prices need not keep down consumption expenditure in housing and other durable goods. The poor statistical results for this period are shown in Table 4.1.

(d) *Summary*

The analysis of private consumption behaviour in Sweden has shown the importance of the financial development, particularly since the middle of the 1970s, i.e. during a period of fairly low growth. The dependence of consumption on financial development brings to the fore the life cycle theory of consumption. In this theory consumption expenditure is explained by current and expected income and by household net worth. The macro-economic consumption function of this theory can be written as follows (Ando and Modigliani, 1963):

$$C_t = \alpha_1 Y_t + \alpha_2 Y_t^e + \alpha_3 A_{t-1}$$

C_t stands for current outlays for non-durable goods and services and the rental value of the stock of service-yielding consumer durable goods, Y_t is current non-property income, Y_t^e is expected annual non-property income and A_{t-1} is household net worth.

In this theory the long-run behaviour of consumption has been particularly emphasized. The studies by Kuznets (1942; 1946) and Goldsmith (1955) had led to the opinion that the long-run consumption function was characterized by equality between the marginal and the average propensities to consume. In order to formulate such a consumption function Ando and Modigliani have made the following assumptions:

1. a constant production growth rate n;

2. an expected production Y^e equal or proportional to current production Y;
3. a reasonably stable rate of return on assets r;
4. a disposable income Y^* equal or proportional to current income Y.

The consumption function implies that the ratio Y^*/A_{t-1} of income to net worth will tend to be a constant h which is related to the parameters of the consumption function by the equation

$$h = \frac{n + \alpha_3 - \alpha r}{1 - \alpha}$$

where $\alpha = \alpha_1 + \alpha_2$. When the ratio Y^*/A_{t-1} is equal to h, the ratio of consumption to income will be a constant given by

$$\frac{Y_t}{A_{t-1}} = \frac{Y_t^* - rA_{t-1}}{A_{t-1}} = h - r$$

$$\frac{C_t}{Y_t} = \frac{Y^* - S_t}{A_{t-1}} \frac{A_{t-1}}{Y_t} = \frac{h - n}{n - r}$$

The long-run consumption function is illustrated graphically in Fig. 4.5.

In the short run the net worth A_{t-1} is given. As Y_t^e is assumed to be equal to Y_t and $\alpha_1 + \alpha_2 = \alpha$, the short-run consumption function can be written

$$C_t = \alpha_3 A_{t-1} + \alpha Y_t$$

The result is a consumption function of the traditional Keynesian character. Graphically it is represented by a linear C^s function with intercept $\alpha_3 A_{t-1}$ and slope α.

In the medium and long term household net worth increases and the short-run C^s curve shifts upwards. A parallel increase in assets and income

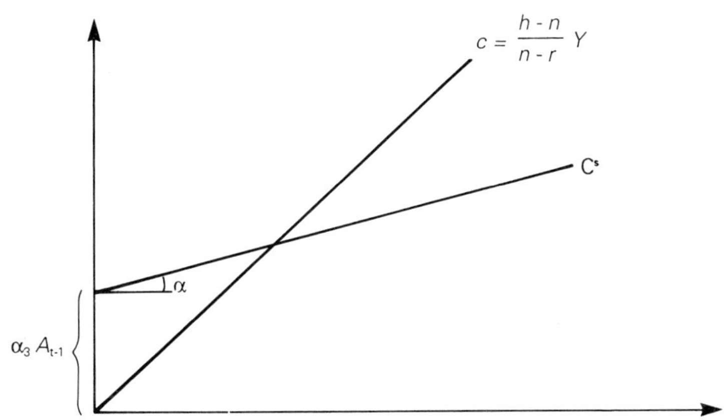

Fig. 4.5 Long- and short-run consumption functions.

characterizes stable growth processes, and such a process was assumed by Ando and Modigliani in their analysis of the long-run consumption function. In Sweden the development during the 1960s and up to the middle of the 1970s resembled such a situation. In this period the financial development did not place any restrictions on private consumption and consumption behaviour cannot be explained in terms of financial variables.

Financial development became important for consumption behaviour when growth stagnated in the period from the middle of the 1970s to the latter part of the 1980s. This was a period when the increasing disequilibrium in international payments and government budget positions led to a strong growth in financial assets in the private sector. It also led to huge movements in both commodity prices and prices of financial assets. In such a situation consumption planning is only possible over short periods, and it may stimulate speculation and a tendency for substitution between different assets and commodities together with a higher sensitivity to expectations about price and interest movements.

It has been shown earlier that the change in the production growth rate in the middle of the 1970s was produced not only by a change in economic policy in the industrial countries, but also by the effects of a long-term structural change in production and income development. This led to short-term changes in income distribution and financial development, which in turn affected the growth and distribution of demand for different types of consumer goods and services. It is natural that such short-term changes will also have consequences for long-run consumption behaviour. This means that it may be more fruitful to start with the short-term consumption function and then study the shifts in this function which may occur in the medium and long term. It is possible that the result will not agree with the hypothesis that in the long run the marginal and average propensities to consume will be equal. There are studies that indicate that the average propensity to consume has declined secularly over the post-war period. The decline during the 1970s has been attributed to the increasing rate of inflation. This development has been explained as the result of increasing uncertainty about future real income, which leads to a tendency to accumulate liquid assets. It has also been argued that in a situation of increasing inflation consumers misperceive relative price changes. A general price rise in consumer goods is perceived as concerning only goods that the consumer happens to demand and makes him curtail his consumer expenditure (Levacic and Rebmann, 1982, p. 223).

Studies have shown that a better fit between actual and predicted consumption is obtained if inflation is included in the consumption function. An example is the following:

$$\Delta_4 C_t = \underset{(0.04)}{0.47}\ \Delta_4 \ln Y_t - \underset{(0.05)}{0.21}\ \Delta_1\Delta_4 \ln Y_t - \underset{(0.02)}{0.10}\ \ln\left(\frac{C}{Y}\right)_{t-4}$$

$$+ \quad 0.01 \underset{(0.003)}{} \ln \Delta_4 D_t^o - \underset{(0.07)}{0.13} \Delta_4 \ln P_t - \underset{(0.15)}{0.28} \Delta_1 \Delta_4 \ln P_t$$

$$R^2 = 0.77 \quad SE = 0.0061$$

where $\Delta_4 C_t$ is the difference between consumption (expenditure on non-durables and services) in the ith quarter of year t and the ith quarter of year $t-1$, $\Delta_4 Y_t$ is the four-quarter difference in personal disposable income, $\Delta_1 \Delta_4 Y_t = \Delta_4 Y_t - \Delta_4 Y_{t-1}$, $(C/Y)_{t-4}$ is the lagged aggregate APC (Average Propensity to Consume), $\Delta_4 D_t^o$ is a dummy variable for 1968 (II) when future indirect tax increases were known, $\Delta_4 P_t$ is the four-quarter difference in inflation and $\Delta_1 \Delta_4 P_t$ is the change in $\Delta_4 P_t$, i.e. a term for acceleration of inflation.

4.2 INVESTMENT DEMAND

4.2.1 The consumption–investment choice

The quotation from Keynes in Chapter 2, section 2.2(b), indicated that investment expenditures are different from other types of expenditures. In *General Theory*, Keynes treats consumption demand as governed by psychological law.[1] This 'law' is stabilizing in the sense that it induces some kind of proportionality into consumption expenditure. In fact, it is quite natural to think of such a law when taking into consideration that the ultimate end of the productive system is human desires which are partly satisfied through the consumption of goods and services.

Investment expenditure is of quite another nature. The investment capital is not demanded for its own sake but as a medium of exchange between present and future consumption. The problem is that production capital is not the only medium which can be used for the transformation of present to future consumption, but it is the only one which increases the total future consumption possibilities of society as a whole. However, from the point of view of the individual it is possible to choose ways of increasing future consumption which imply a redistribution of future relative to present consumption among individuals. The fundamental problem in the theory of investment is partly to analyse the choice between consumption and savings and partly to analyse the reasons why individuals choose one type of investment rather than another. From the point of view of society as a whole, as great a part of savings resources as possible should be directed to productive resources. However, this could conflict with individual motives

[1]The fundamental psychological law, upon which we are entitled to depend with great confidence both *a priori* from our knowledge of human nature and from the detailed facts of experience, is that men are disposed, as a rule on the average, to increase their consumption as their income increases, but not by as much as the increase in their income (Keynes, 1936, p. 96).

for selecting a portfolio of assets to ensure future prosperity when individual uncertainty and liquidity restrictions are taken into consideration.

We mentioned in Chapter 1 that liquidity restrictions are not unique to a monetary economy but are also possible in a barter economy. A liquidity restriction indicates that the agent is forced to have certain goods or services irrespective of future consequences. As we mentioned earlier, this is equivalent to lexicographic preferences. The closer an agent is to the necessary goods and services the higher is the risk premium which reaches infinity at the lexicographic minimum. This means that the agent in a neighbourhood chooses the most liquid form of saving, if any.

Keynes discusses motives for holding liquidity. He mentions three different ones: (a) the transaction motive, (b) the precautionary motive and (c) the speculative motive. We could interpret the precautionary motive as security for future basic consumption.

The usual picture of investment in equilibrium analysis is that it represents future consumption and thereby stays in a substitutive relationship to current consumption. Efficient investment of withdrawn current consumption increases the possibilities for future consumption. We can represent the opportunity cost of investments in the diagram shown in Fig. 4.6. Trajectories $C_1(t)$ and $C_2(t)$ represents different growth rates of future consumption up to the time horizon T. The withdrawn consumption A, which is invested in some appropriate project, is the opportunity cost of the future consumption B. If the agent demands a greater quantity B than the withdrawn quantity A, the difference will reflect the time preference of the agent. There are usually three arguments in favour of a positive time preference. The basic argument is the inevitability of death. We must enjoy life when it is possible. The second type of argument refers to the general risk, or rather uncertainty, which is inherent in expected future consumption. The third type of argument is of another character. It refers to an assumption of a general economic growth which will increase future

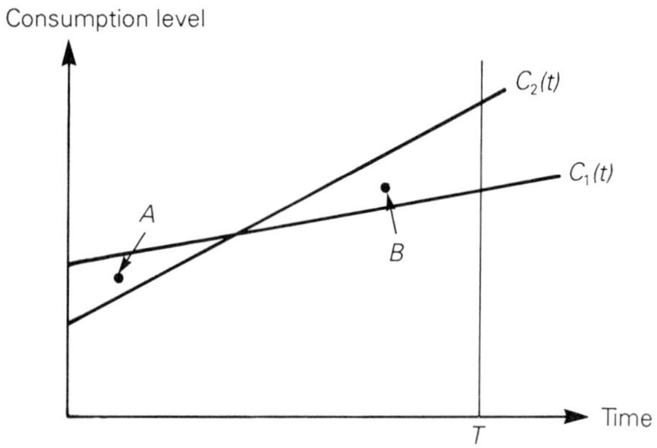

Fig. 4.6 Opportunity costs of investments.

consumption possibilities. If the agent takes a pessimistic view of future propects this will not necessarily decrease the time preference because he may then switch to the two former arguments which in this case may increase the time preference. The effect of the third argument is not obvious but depends on the wealth of the agent, social liabilities and expected future liquidity restrictions. An increase in expected future growth may, for example, decrease the level of savings but induce the agent to choose a less liquid portfolio with better prospects of future yield.

It is obvious that Fig. 4.6 could equally well represent the savings–consumption choice instead of the investment–consumption choice. If the economy is running along an equilibrium growth path where we have perfect information about the future of perfect future markets, i.e. uniform information about possible outcomes and uniform probability distributions, the two choices will coincide. In a disequilibrium world where the agents and the economy are normally in a state of uncertainty, recalling our discussions in Chapter 1, there will normally be a difference between the two types of decisions from the individual agent's point of view.

The word 'savings' emphasizes the transmission of current resources into future usage, while the word 'investment' emphasizes the growth aspect. However, this difference is only relevant to the individual agent. For the economy as a whole the word 'savings' normally refers to the transmission of productive resources and will then coincide with investment.

Agents can use money as well as goods for savings purposes. The central aspect here is the liquidity of the object in question. If we choose a good as a store of value we are dependent on the development of the demand for that particular good and the rate of depreciation. We may refer to the central condition in general equilibrium theory that the elements of the initial endownment vector must be present in at least one agent's preference relation (e.g. Debreu, 1974; Mas-Colell, 1985). If we disregard inflation, the medium of exchange, i.e. money, is the only store of value which is independent of variations in demand and will therefore minimize the risk. We refer here to a closed economy; in an economy dependent on foreign markets we must take currency risks into consideration. We can conclude that increasing foreign trade will decrease the liquidity of the local currency, if by the word 'liquidity' we mean the minimum dependence of variations in demand. When the agent chooses money, or liquid securities, as a store of wealth for the future, the absolute future value will be a positive function of the general growth of the economy but the value of the wealth relative to his fellow citizens will be a negative function. That is, when there is positive growth the relative wealth distribution will normally be in favour of risky assets, while when there is negative growth the relative wealth distribution will be in favour of liquid assets. Consequently we find that there is no obvious connection between individual savings and individual investment in productive assets.

We can now ask whether the total investment in an economy will be the sum of individual investments. The problem here is whether the total growth of resources is the sum of the growth of the individual agents. This question is fundamental to the understanding of the difference between equilibrium and disequilibrium. When the economy experiences an equilibrium growth, if this is empirically possible, the price vector in the economy will maintain the distribution between the agents. In disequilibrium, however, the agents trade at false prices by definition. Trading at false prices will, from a theoretical point of view, change the distribution. This will imply that there are two sources of individual growth. On the one hand the individual could invest in productive resources, which will increase the resources of the total economy. On the other hand he could invest in redistributions of future resources. The most common example of the latter is when an agent utilizes the binding liquidity restrictions of his fellow agents to increase his future wealth. This was explicitly prohibited by Thomas Aquino in a society built on divine laws. To request interest payments on lent capital was to utilize the misfortune of a brother.

Let us now try to incorporate these considerations into the neoclassical theory of investment. We normally divide the consumption–investment decision into two subquestions. First, the individual chooses between current and future consumption and, secondly, the agent chooses a suitable portfolio for his savings. In this approach there is a complete substitutional relationship between current and future consumption. However, if we introduce the concept of lexicographic preferences we have to modify our attitude towards substitution. As we discussed in Chapter 1, lexicographic preferences occur when the agent is forced to have a certain quantity of some good irrespective of its relative price. We can interpret this in a monetary economy as binding liquidity restrictions. The agent knows in general that he is going to face future liquidity restrictions but it is only the current restrictions which are binding. However, the presence of future liquidity restrictions will influence both the consumption–saving choice and the portfolio decision. If we allow for the possibility of short-term borrowing we shall easily arrive at a situation where the fulfilment of current liquidity restrictions may aggravate the future liquidity position.

Current consumption will influence future consumption in two ways. On the one hand the level and distribution of current consumption will affect future consumption by forming 'consumption habits'.[2] These habits include both habits in the usual sense of the word, which means that the desired standard of living will change continuously, and habits acquired through consumption technology formed in particular by the acquired stock of durable and semi-durable goods. However, the current consumption will

[2]For a discussion of consumption habits from the individual's point of view see for example Pollack (1970).

alter future liquidity restrictions as a result of the way that it is financed. An increasing ratio of debt to income will deteriorate the future liquidity position provided that there is no corresponding liquid wealth. An increase in the ratio of durable goods to total consumption may then have an effect on current liquidity restrictions and thereby influence the elasticities of the non-durable goods which are primarily financed out of current income flow.

A comparative statics formulation of the consumption–saving choice when current and future consumption restrictions are present could have the form

$$\max U = U(C^{c}, C^{f}) \tag{4.1}$$

subject to

$$C^{c} > C_{0}^{c} \quad C^{f} > C_{0}^{f} \tag{4.1a}$$

$$P^{e}C^{c} + m_{c} + \left(\frac{1}{r^{e}}\right) \times a = wL + \left(\frac{1}{r^{e}}\right) \bar{a} + \bar{m} + d^{cx} i \tag{4.1b}$$

$$P^{f}C^{f} = W^{f}L^{f} + \bar{m}_{f} + \left(\frac{1}{r^{e}}\right) \times a \tag{4.1c}$$

The agent maximizes his utility which is a function of current and future consumption. The first two restrictions are the lexicographic element in current and future consumption. Restriction (4.1b) is the current budget, which on the left-hand side consists of current consumption outlay, demand for more balances and demand for risky assets. The price of the risky asset is the inverse of the rate of return and a is the number of perpetuities yielding one unit of money value. On the right-hand side we have wage income and initial balances of risky assets, money balances and debt on which services have to be paid. Restriction (4.1c) is the expected future budget restriction including expected wages, prices, planned labour supply, planned consumption, money balances and finally risky assets brought from the current period.

Without going deeper into the mathematical details of this rather complex optimization problem we can illustrate the effects of the two lexicographic restrictions as shown in Fig. 4.7.[3] The slope of the budget line represents the ratio of expected to current income. A steeper slope indicates an increase in expected income. In Fig. 4.7(a) we analyse the effect of the restriction on current income. Points 2 and 3 are the unconstrained optima given the respective expectations. As a consequence of the restriction the current consumption will become less than in the unconstrained optimum and thereby reduce savings to points 2′ and 3′ respectively. The introduction of a restriction on the future consumption level in Fig. 4.7(b) will

[3]For a complete discussion of the choice between current and future consumption under a normal preference structure see Morishima (1973).

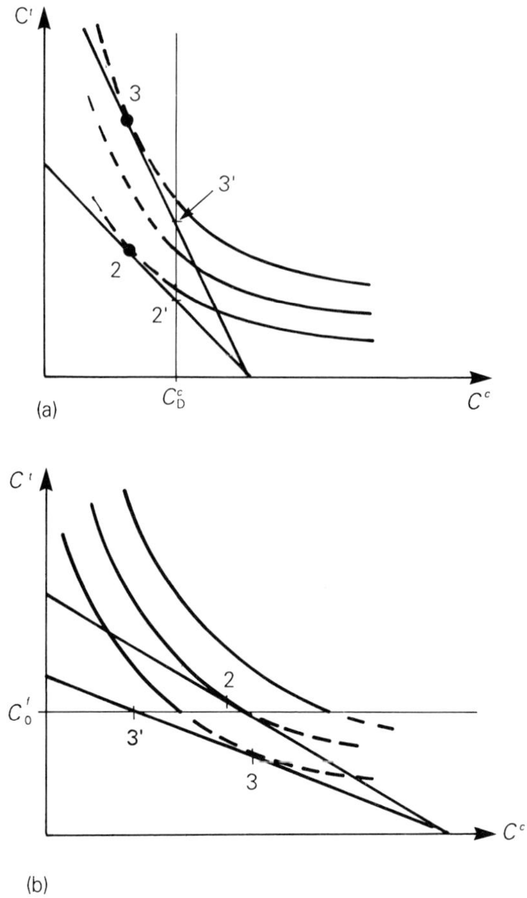

Fig. 4.7 Lexicographic restrictions on current and on future consumption.

decrease the consumption relative to the unconstrained optima when expectation of future income deteriorates, and in this case savings will decrease less.

Analysis of the choice between a risky asset and liquidity is rather straightforward if we have variations in expectations, but we shall return to this choice in discussing portfolio selection. If the two situations are characteristic of the total economy, we would have the interesting situation that the restrictions on current consumption and thereby liquid resources made the increase in savings less than the increase in expected future income, whereas in the second case the restrictions on future consumption cause a decreasing current demand (which will probably cause a further effect on expectations) instead of a counteracting increase or at least a stable current demand. In the first situation the willingness to invest in risky assets is high but savings will be suboptimal. In the second case there will be an increase in saving but an increasing liquidity preference which will worsen the financial conditions for risky capital.

In the above discussions it is essential to realize that the Keynesian problem of information deficiency between savers and investors and the demand for liquidity is not a problem exclusively for monetary economies. We can describe the problem in terms of liquidity restrictions which are normally disregarded when we assume perfect future markets, homogeneous goods, uniform information etc. In a barter economy we could also consider liquid and non-liquid goods. An agent is liquid in a barter economy if he possesses a stock of goods which occur frequently in his fellow agents' utility functions and the demand is inelastic with respect to price and the supply is stable. This means that in a monetary economy we can counteract increasing uncertainty by collecting the medium of exchange, but we can also choose a strategy which is rational in a barter economy and which is also available in a monetary economy to increase control of goods as is frequently demanded, i.e. to reduce competition in important markets. It also follows from the above discussion that when there are restrictions of the types we have discussed there are no 'automatic' stabilizing forces on the financial side to counteract increasing real uncertainty.

The traditional view of the consumption–investment choice originates from Markowitz (1959) and was further developed by Fama (1970) and Merton (1973). We shall briefly summarize their approach and make some comments from a 'disequilibrium' point of view. The central point in their approach is to try to find suitable conditions which make it possible to prove that a decision, which takes many periods into consideration and where the agent tries to optimize a trajectory over these periods, can be reduced to a one-period choice. The advantage of this is obvious. If it is possible and the conditions are reasonably realistic from an empirical point of view, the agent could avoid the multi-period calculation and just consider the optimal choice for the next period. Fama draws the following conclusions:

> In sum, assuming only that markets for consumption goods and portfolio assets are perfect and that the consumer is risk averse in the sense that his utility function is strictly concave, it has been shown that though he faces a multiperiod problem, in his consumption–investment decision for any period the consumer's behaviour is indistinguishable from that of a risk averter who has a one-period horizon. It was then shown how this result can be used to provide a multiperiod setting for the more detailed hypotheses about risk averse behaviour that are traditionally derived in a one-period framework. (Fama, 1970)

Going back to Arrow and Debreu, general equilibrium can be proved if the agent possesses perfect information about the future or if perfect future markets exist. The proof by Fama, developed further by Markowitz, is obvious in the line of the latter condition, in this case completed using von Neuman–Morgenstern utility functions. We shall first discuss the proof by Fama and then comment on the extension by Merton.

First, it is assumed that the agent's incomes are capital incomes.[4] A cardinal utility function is then assumed for period $\tau + 1$:

$$U_{\tau+1}(C_{\tau+1}|\beta_{\tau+1}) = U_{\tau+1}(c_{1-k}, \ldots, c, \ldots, c_{\tau+1}|\beta_{\tau+1})$$

where $c_{1-k} \ldots c_1$ is the historical consumption until current time and $c_{\tau+1}$ is some form of bequest. The utility function is additive with respect to consumption in the different periods. The parameter β_τ represents the state of the world at time period t. The consumer seeks to maximize the expected utility of his lifetime consumption.

The individual agent has to solve the following problem:

$$U_\tau(C_{\tau-1}w_\tau\beta_\tau) = \max_{c_\tau, H_{\beta_\tau}} \int_{\tau+1} U_{\tau+1}\ (C_\tau,\ H\cdot'|\beta_{\tau+1}\ \mathrm{d}F_{\beta_\tau}(\beta_{\tau+1}) \qquad (4.2)$$

subject to the constraints

$$0 \leqslant c_\tau \leqslant w_\tau \quad H_{\beta_\tau} i' = w_\tau - c_\tau \quad H_{\beta_\tau} \geqslant 0$$

The problem is solved using dynamic programming — a backward optimization. The agent decides the optimum values for the ultimate period and then, using these as restrictions, he chooses the optimum values for the immediate preceding period. Then, using the latter values as restrictions for the next immediate preceding period, he continues until the optimum values for the current period are found.

In addition to the assumption of additivity it is also assumed that the consumption at period t depends only on the state variable vector and not on preceding states of the world, which means that the consumption in one period is independent of the consumption in preceding periods. Under these assumptions it is shown that the multi-period choice is equivalent to the one-period choice.

Merton (1973) developed expression (4.2) more explicity along the lines of the life cycle hypothesis. A random consumer acts so as to

$$\max E_0\left\{\int_0^{T^k} U^k[c^k(s), s]\,\mathrm{d}s + B^k[W^k(T^k), T^k\right\} \qquad (4.3)$$

where E_0 is the expectation operator conditional on the current value of his wealth, $W^k(0) = W^k$ are the state variables of the investment opportunity set and T^k is the distribution for his age of death, independent of the

[4]Merton (1973, p. 875) admits that since there are no institutional possibilities of issuing shares against future wage incomes, the introduction of wage income will cause systematic effects on the portfolio and consumption.

investment outcomes. The utility function has the same characteristics as in the above model and B^k is the 'bequest' or utility of the terminal wealth $W^k(T^k)$.

Merton also defines a perfect capital market structure which we will discuss in more detail in Chapter 5 on portfolio selection. Important assumptions are that the markets are always in equilibrium and that short sales of all assets with full use of the proceeds are allowed. Moreover, the expected rate of return per unit of time is defined by

$$\alpha \equiv \frac{E_t\{[P(t+h) - P(t)]/P(t)\}}{h} \tag{4.4}$$

and the variance

$$\sigma^2 \equiv \frac{E_t\{[P(t+h) - P(t)]/[P(t) - \alpha h]^2\}}{h}$$

which is a measure of risk. The interval h is sufficiently small to guarantee continuity and the changes in the state variables are local. The dynamics of the asset prices are purely random processes and fulfil the conditions of a Wiener process with a given mean, i.e. the wealth is given at every moment of time.[5]

During these conditions we shall have a wealth accumulation process only, dependent on the choice between current and future consumption. The accumulation process will not be interrupted by windfalls in either direction. The analysis by Merton explicitly details the conditions for the asset markets which must be fulfilled if the Fama analysis is to be correct.

The crucial assumptions in the consumption–investment analysis are as follows: (a) consumption at period t depends only on the state of the world at t; (b) the consumption at period t is independent of consumption in earlier and later periods; (c) 'backward' optimization; (d) the wealth is given in each period and is unaffected by market operations. Compared with our earlier discussion this implies that the current consumption is a residual given the time preference; future consumption incurs restrictions on current consumption. Furthermore, the behaviour of market prices is random in every period which means that the agent just needs information about the variance relative to an expected mean. According to the discussion of risk and uncertainty in Chapter 1, the economy discussed by Fama and Merton in the capital asset pricing model is obviously of type 1 for the economy as a whole and type 1 for the individual with respect to our classification in Chapter 1, p. 6. That is, all individual agents have identical subjective probability distributions over the space of outcomes. When we take the *ex ante–ex post* discussion into consideration we find an individual whose *ex*

[5] A Wiener process is a real-valued Gaussian process such that (a) it has independent increments and (b) $w(s+t)\ldots w(s)$ is $N(O, Q^2 \cdot t)$ for all $s_1 t \geqslant 0$ where $Q^2 > 0$. See, for example, Grimment and Stirzaker (1982, p. 257).

ante plans are unaffected by the result of comparisons *ex post* between earlier *ex ante* plans.

Models of this type are perhaps suitable for describing the behaviour of short-run security markets, but in their assumptions they exclude physical capital which has a long time span and which is inflexible with respect to technical properties.

4.2.2 Investment decisions by non-financial enterprises

In the transactions table (Fig. 1.10) we register real investments in the B matrix and financial investments in the columns of the F matrix. We can interpret the rows of the Π and F matrices as the budget restriction for investment expenditure. Our interest here mainly concerns investments in productive capital made by non-financial enterprises. We shall concentrate on this sector and its aggregate investment behaviour.

Fundamental to modern industrial economies is the division between owners and entrepreneurs. In general owners are anonymous shareholders, individuals or organizations with limited influence and less operative and financial responsibility. The entrepreneurs, in contrast, have no major economic interest in the firm in general but have the major operational responsibility for running the firm. The fundamental effect of this division with respect to our earlier discussion is the separation of the owners' time preference from the organization's time preference. This will imply that the short-run finance of investment is effectively independent of the valuation of the firm's equity on the second-hand market. If the internal financing ability is good the firm increases the time horizon of its decisions without reference to the owners' situation. This will of course lead to conflict if the difference between the time horizons of managers and owners is too great. We shall return to this point in Chapter 5.

(a) *The investment decision*

A fundamental condition for the stability of an equilibrium in general equilibrium theory is gross substitution. The investments of non-financial enterprises in physical capital are registered in the columns of the B matrix and their financial investments are registered in the F matrix. Our main aim in this section is to discuss the portfolio decisions of non-financial enterprises and particularly whether their investments in physical capital are likely to be treated as arbitrary assets. We are also concerned with the question of the stability of the financial market and whether price formation in this market will counteract sudden changes in expectations of the future yield of physical capital based on imperfect information. Following Keynes's discussion in *General Theory* (Keynes, 1973, Chapters 5 and 12), we could illustrate the problem in the following way.

Assume that the demand of an economy has been stable up to time t_0. The capital stock is stabilized relative to this demand, and so are investments if we think of some steady state path. However, let us assume that the long-term expectation of demand by the entrepreneurs is changed at time t_0 and the demand is expected to increase to a higher level which is true in an 'objective' sense. The effect of this change will be that investment demand will increase to a higher level than the optimum in the long-term perspective during the expansion phase of the capital stock. By then, the aggregate demand during this period will be higher than the long-term expectations. If the long-term expectations do not change during this expansion phase but are constant, there will be a converging process towards the higher stable long-term demand. We illustrate the process in Fig. 4.8. However, if the expectations are based on development during the immediate past, the convergence is less certain and we would experience some kind of destabilizing accelerator process, at best oscillating within some limits or leading towards some other long-term growth path on a lower level. We indicate these shifts in expectations by the broken lines.

The question is whether the financial markets are able to counteract changes in the subjective risk of physical capital given a long-run growth path. By counteraction we mean that the prices of alternative assets to physical capital behave in such a way that the increase in risk is met by a decrease in the price of physical assets or an increase in the price of the substitutes which will offset the decrease in the demand for physical capital. Empirically firms are generally strongly involved in financial markets. Figure 4.9 shows shares of financial capital to total capital during the period 1977–84 found in a study by the Government Board for Industry of Sweden (SIND, 1986).

When the Stockholm School introduced the *ex ante – ex post* distinction during the 1930s it was an important contribution to disequilibrium

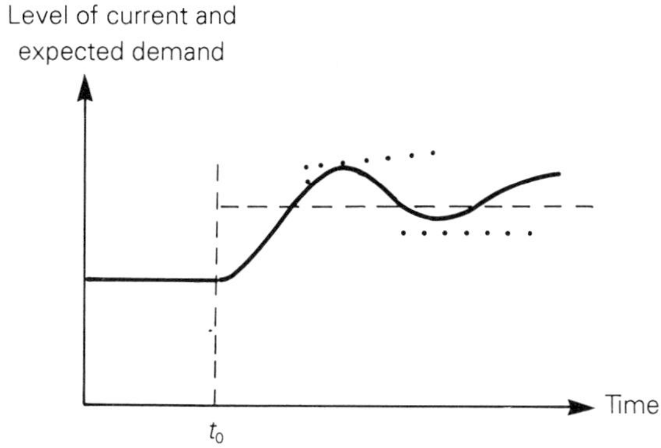

Fig. 4.8 Illustration of fixed and flexible long-run expectations.

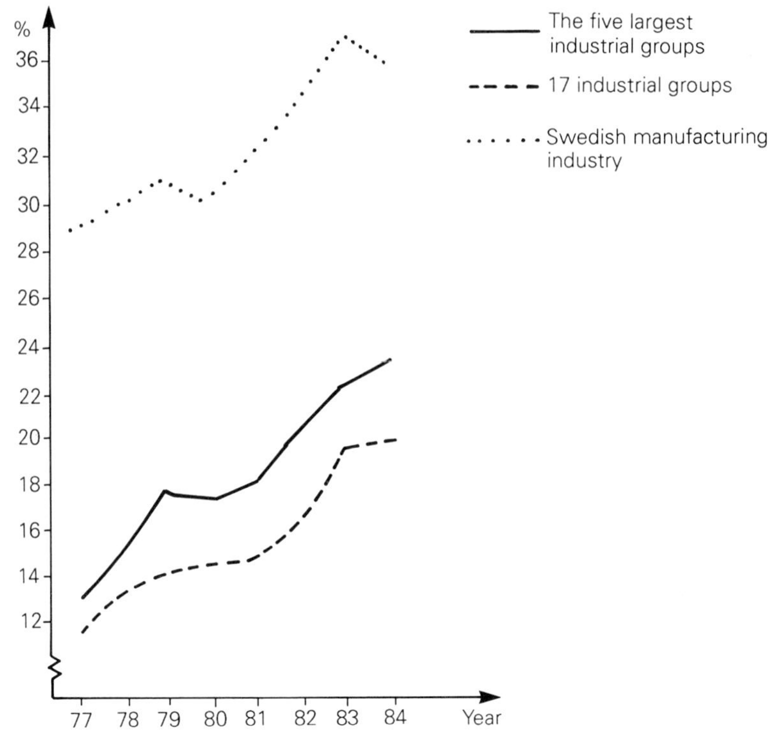

Fig. 4.9 Relation between financial and total capital in Swedish industries.

analysis. In general equilibrium analysis the agent acts under a perfect future market and a subjective risk distribution based on full information about future possible outcomes. Furthermore, the prices will be equilibrium prices either at one time for all future deliveries or at each time under the assumption of perfect flexibility of the relevant factors of production. Given the standard assumptions of a rational agent, the prices at each time will be unique and contain all the necessary information about the preference structure for the given distribution of productive resources. If there is a continuum of consecutive markets the economy will follow an equilibrium trajectory at a given relative distribution of wealth and income among the agents.

In Chapter 1 we discussed the concepts of risk and uncertainty, and we stated that the normal situation in the empirical world is that agents have different opinions of the space of possible outcome and in the worst case the union of the individually perceived spaces of outcome is a genuine subspace of the universe. Earlier in this chapter we discussed the effects of insignificant flexibility of factors of production combined with complementarities.

The effect of such 'imperfections' is that information contained in prices will be obscured and the one-dimensional image of the multi-dimensional preference space will not be unique. However, this is the world of the *ex ante – ex post* distinction. The agent, in our case the investor, makes a decision based on the relevant information actually available at the moment of decision. If the decision is irreversible, which is the case when there are fixed factors of production and to some extent fixed relations between different factors of production, and has consequences for future incomes and future liquidity restrictions, the agent will face an *ex post* situation which is different from that expected *ex ante*. From an *ex post* point of view the *ex ante* action could indicate false trading relative to the *ex post* information and restrictions, and thereby induces redistribution of wealth.

This situation is covered to some extent in the stability analysis of the general equilibrium theory. The *ex ante – ex post* discussion is similar to the stability of equilibrium problem apart form the fact that the latter assumes equilibrium at the outset.

The stability problem is central to the theory of investment. According to the quotation from Keynes in Chapter 2, section 2.2(b), an increase in saving will withdraw current consumption and perhaps induce investors to decrease their investment activities. This question will of course increase in importance if we add the problem of future liquidity restrictions. However, we could imagine a financial market where these effects are counteracted.

To state the problem according to our transactions table, we may ask whether price formation in the security market will counteract increasing subjective risk to physical capital caused by inefficient information. The difference between *ex ante* and *ex post* could be interpreted as exogenous shocks created by inefficient information but still in some way external to the price system at a particular time. In such cases counteracting price movements on the security market could restore the equilibrium, provided that the equilibrium is on some feasible long-term growth path.

The modern analysis of the stability of both temporary and general equilibria is based on the work of Hicks (1939). His basic argument is that comparative static analysis has no meaning if dynamic stability cannot be proved for the economy in question. He states:

> The laws of change of the price-system, like the laws of change of individual demand, have to be derived from stability conditions. We first examine what conditions are necessary in order that a given equilibrium system should be stable; then we make an assumption of regularity, that positions in the neighborhood of the equilibrium position will be stable also; and thence we deduce rules about the way in which the price system will react to changes in tastes and resources.
>
> What is meant by stability in exchange? In order that equilibrium should be stable, it is necessary that a slight movement away from the

> equilibrium position should set up forces tending to restore equilibrium. (Hicks, 1939, Chapter 5)

Hicks defines two concepts of stability. First there is 'imperfect stability'. A market is imperfectly stable if a fall in the price of one good implies an excess demand for this good after all other markets are cleared. However, a market is perfectly stable if a fall in the price of one good implies excess demand for the good in question when a given subset of other markets are cleared and the complementary equilibrium prices are constant.

The concept of perfect stability was utilized by Metzler (1945) who showed it to be equivalent to the condition of gross substitution. Metzler also showed that the condition of gross substitution is necessary if the stability is to be independent of time convergence. This condition is fundamental to the analysis of portfolio selection (see, for example, Tobin, 1980, 1982; Merton, 1982). In portfolio analysis, however, the expectations of future prices play an important role; a natural extension of the stability analysis is therefore the inclusion of expected as well as current prices. [6]

One of the earlier interesting discussions of stability in relation to expectations is that of Enthoven (1956), which is extended by Enthoven and Arrow (1956), concerning extrapolative expectations. A short survey of the results is given below.

Assume the excess demand function

$$x_i = X_i(p_1, \ldots, p_n, p'_i); \quad i = 1, \ldots, n \tag{4.5}$$

where p_i is the current price and p'_i is the expected price. In equilibrium we have $x_i = 0$ and $p_i = p'_i$. Define the expectation formation (extrapolative) as

$$p'_i = p_i + \zeta \dot{p}_i \quad \zeta > 0 \tag{4.6}$$

where

$$\dot{p}_i = \frac{dp_i}{dt} = K_i x_i \quad K_i > 0$$

Enthoven and Arrow show that, if the economy fulfils the gross substitution condition in the absence of expectations, the condition for stability for one market when extrapolative expectations of the kind defined above are present is as follows (the result can be generalized to the whole economy and this is done in the original paper):

$$\frac{1}{k_i} > b_i \zeta_i \tag{4.7}$$

where $b_i = \delta x_i / \delta p'_i$. The condition implies an inverse relationship between the sensibility of expected prices and the rate of adjustment of prices. This

[6] Note that trading only occurs at equilibrium prices.

result is of great interest when prices are used for information about qualitative aspects of the goods in question. If the result is related to the goods market we have to assume some kind of Veblen goods, but in the analysis of financial markets where information is not uniform the prices of different assets typically have a double role. In such markets it is not unreasonable to assume extrapolative expectations during certain periods.

Arrow and Nerlove (1958) analyse stability with respect to adaptive expectations. They define the expectation formation as a function of the difference between current prices and the expected prices for the current period. This problem is comparable to the *ex ante–ex post* problem. The expectation function is defined as follows:

$$p'_i = \beta_i(p_i - p'_i) \tag{4.8}$$

where $\beta_i > 0$. The solution of the differential equation is

$$p'_i = p'_i(0) \times \exp(\beta it) \int_0^t \beta_i p_i(\tau) \exp(\beta i \tau)\, \mathrm{d}\tau \tag{4.9}$$

Provided that the initial period is sufficiently remote, the expected price will be equal to the weighted average of the historical prices. However, if the initial period approaches the current period, the adaptive expectation will approach the extrapolative expectation.

Arrow and Nerlove define two matrices A and B:

$$A = \begin{bmatrix} a_{11} & \cdots & a_{1n} \\ a_{n1} & \cdots & a_{nn} \end{bmatrix} \qquad B = \begin{bmatrix} b_{11} & \cdots & b_{1n} \\ b_{n1} & \cdots & b_{nn} \end{bmatrix} \tag{4.10}$$

where $a_{ij} = \delta x_i / \delta p_j$ and $b_{ij} = \delta x_i / \delta p'_j$. Under the above conditions the economy will be stable if the matrix $A + B$ satisfies the gross substitution requirement. This implies either that if the reactions of the agents with respect to expected prices dominate the reactions with respect to current prices the goods must be substitutes in expected prices, or, which is the more common interpretation, if the goods are gross substitutes in current prices the reactions to those prices must dominate the reactions to the expected prices.

We postpone the general discussion of portfolio selection to the next chapter and focus our attention on investments in physical capital and their relations with other types of assets. It is important to realize that existing capital stock in its marketable form, equities, is an alternative asset to financial assets for new capital. The fundamental difference between new physical capital and all other assets is that its output price is independent of expectations of its future yield and depends only on its current costs of production. The standard decision problem for physical capital is written in the following form:

$$pI \leqslant \int \mathrm{NC}(t) \exp(-rt)\, \mathrm{d}t \tag{4.11}$$

where P is the price of goods, I is the quantity of investment, NC is the expected net inflow of cash and r is the discount rate. Assuming that the expected net inflow of cash is based on current demand, prices and production costs and an expected growth factor, we can transform the problem to the following form:

$$I \leqslant \frac{\mathrm{NC}}{P} \int \exp(g - r)t \, \mathrm{d}t \tag{4.12}$$

where g is expected growth. If we disregard the 'less than or equal to' sign and interpret the last equation as the demand function for investment goods, we have the classical problem discussed by Wicksell (1916). Investments are affected by the difference between the market rate of interest and the expected real growth and the level of current prices of investment goods.

The last expression is very similar to the famous Tobin q. However, the difference is that Tobin's q is defined as the ratio of the market valuation of equities to the replacement costs of capital. In his Nobel Lecture in 1981, Tobin (1982) specified the investment function

$$I_t = q_t^k K_{t-1} f(q_t^k) = q_t^k K_t \tag{4.13}$$

where

$$\Delta k_t + \delta K_{t-1} = 0$$
$$f(1) = g$$
$$f'(q_t^k) \geqslant 0$$
$$0 < f^{-1}(-\delta) < 1$$

K_{t-1} is the current capital stock during period t, valued at replacement costs. The factors δ and g are the depreciation rate and the natural rate of growth repectively. Tobin's formulation of the investment function is based on a general equilibrium approach where the market valuation of equities is equivalent to the marginal efficiency of capital. Tobin then proceeded to define the market condition for the equity market:

$$A_t^K(\cdot) - q_t^k K_{t-1} = q_t^k K_{t-1} f(q_t^k) \tag{4.14}$$

The arguments in the demand function for equities $A_t^k(\cdot)$ include the expected returns for all assets. The left-hand side expresses the demand for new capital while the right-hand side expresses the supply of new assets.

Given gross substitution between the different assets, as Tobin assumes, we can expect counteracting price movements on other assets when the expected return on equities deteriorates. This will decrease the elasticity of the total demand for equities $A_t^K(\cdot)$. It is then easy to see that a decrease in q will have a greater effect on the right-hand side of the demand equation, by the conditions of the function $f(Q)$, than on the left-hand side and therefore will create an excess demand for new capital which will counteract the drop in investments created by the deterioration in expected return on new capital. However, this must imply that the market valuation of equities will influence the market price of new investment because the expected cash flow is given. The replacement cost of capital is assumed given in the short run, which is probable since the output price of new capital is determined by production conditions.

The equation for demand of new capital, when the basic argument is Tobin's q, is obviously an *a priori* assumption of stability concerning demand for investment goods and thereby the total demand with respect to changes in expectations by investors. From our point of view this demand expression is not very suitable. In an *ex ante* situation it is obvious that an alternative to new capital is the capital that already exists. The major difference between new and old capital is price formation. The output price of new capital is determined by production conditions and, according to our considerations of costs and productivity in physical production, we can expect rather sticky prices; therefore changes of the demand will mainly cause quantity adjustments. This will mean that, according to expression (4.12), the only factor which can counteract deteriorating expectations of future growth is decreasing rates of interest. However, price formation on both sides of the second-hand equity market is based on expectations of future returns and future prices.

Let us assume an *ex post* situation at time t_0 which is different from the *ex ante* expectation at time t_{-1} in a negative way and thereby affects expectations about the future. We further assume four kinds of assets, new capital, equities on the second-hand market, bonds with a contractual stream of future cash and cash.

(i) Market for new capital. We assume that the supply of new capital is perfectly elastic during the current period. The demand for new capital is given by expression (4.12). The equilibrium condition implies that the realized demand is given by the same expression for fixed price and fixed expected growth and by the rate of interest which is dependent on other markets. Note that we just change the expected growth rate, and not the current cash flow NC. We shall relax this condition later. Inserting the given supply price into the demand equation we have the realized investments

$$I = \left(\frac{\overline{\mathrm{NC}}}{\bar{\mathrm{P}}}\right) \times \int_0^{\infty} \exp(\bar{g} - r)\,\mathrm{d}t \qquad (4.15)$$

where the bars indicate given values.

(ii) Market for equities. The market for equities is slightly more complicated. In standard theory of portfolio selection one of the basic assumptions is that short sales of all securities, with full use of proceeds, is allowed. This

assumption limits the supply of shares to newly issued shares only, thus making the supply side of the equity market passive and 'sterile'. From empirical experience, however, it is clear that the supply of equities depends as much as the demand side on expectations of future returns and prices. A natural assumption is that new and old equities are perfect substitutes. The supply of equities on the second-hand market depends mainly on two factors. First, there is the possibility of making a capital gain through immediate realization of an equity. Second, there is the possibility of future gains through changes in the price. These are generally equal to changes in the present value of the expected future flow of yields. A general formulation of the supply function of equities, if we disregard the need to sell to cover a necessary liquidity demand, could be

$$E_0^s = \Psi\left(\frac{p_0}{\bar{p}_{-t}}, \frac{p_0}{p_0^e}\right) \tag{4.16}$$

where p_0 is the current price, $\bar{p}_{-t}$ is the price at which the equity was bought, set equal to current price, and p_0^e is the expected price which is equal to the present value of the expected stream of cash, and the condition $\delta p_0^e/\delta r < 0$ holds. However, the demand side depends only on the relation between the current price and the present value of the expected future yields which is assumed to have a constant relation with the yield of total capital. We can write

$$E^d = \mu\left(\frac{p_0^e}{p_0}\right) \tag{4.17}$$

If we assume uniform information on both sides, it is reasonable to transform the two equations to the following expressions:

$$E_0^s = \Psi\left(\frac{p_{-t}}{p_0^e}\right) \tag{4.18}$$

$$E_0^d = \mu(p_0^e) \tag{4.19}$$

The stability of the market depends on the relative elasticity of demand and supply. From the supply function we can see that the elasticity depends on the relation between expected price and purchase price and the effect of the rate of interest on expected prices. The demand equation depends only on the expected price, which is affected by the rate of interest. It is obvious from the two expressions that the instability increases when the ratio in the supply function decreases but we have a natural dynamic equilibrium when the ratio is unity, which means that

$$p_0^e = p_0 = p_{-t} \tag{4.20}$$

The condition is that there is no need to sell because of binding liquidity restrictions.

To discuss the dynamic properties we define the expected prices as

$$p_0^e = \text{NC} \times \exp[(g - r)t] \tag{4.21}$$

For simplicity we set NC = 1. We define the price change as a function of the excess demand:

$$\dot{p} = \gamma(E_0^d - E_0^s) \tag{4.22}$$

Simplifying the notation, we can write the expected prices as $p^e = \exp[(g - r)t]$, i.e. a function of expected growth and the market rate of interest. We also linearize around equilibrium:

$$\dot{p} = a\exp[(g - r)t] - bp_0\exp[(r - g)t] \tag{4.23}$$

Expression (4.22) shows that when the economy is moving along a steady-state growth the equilibrium price will depend on the relative elasticity of demand and supply. If we take an ordinary Wicksellian view that in steady-state growth the market rate of interest must coincide with the growth rate, i.e. $g = r$, we will have the equilibrium price:

$$\hat{p} = a/b$$

The solution of the non-homogenous function in (4.23) is rather complicated, and hence it is easier to discuss the relations in terms of the derivative.

From expression (4.22) it is easy to see that the excess demand will be greater for greater differences between current and expected price. The 'active' side here is the supply side, while the demand side defines the new equilibrium. This means that, given the elasticities a and b, the time of convergence to the new equilibrium determined by the demand price will be independent of the difference between the current and expected price. Furthermore, the time of convergence will depend only on the absolute value of the sum of elasticities of supply and demand. It is obvious that the only means of affecting the change in expectation of future growth is an equal change in the rate of interest in the same direction.[7]

From the analysis of the market for new capital and the market for equities we can conclude that they are both dependent on the rate of interest but, contrary to Tobin's analysis, they are independent of each other. Equities adjust quickly to the expected rate of growth and thereby maintain their yield. In the market for investment goods, however, prices are determined by production conditions and we shall therefore have quantitative adjustments, which is the traditional Keynesian view. The only case when we may expect influence from the equity market on the market

[7] The solution is (see, for example, Sanchez 1968):

$$p(t) = \exp\left(\frac{a}{g - r}\{\exp[(g - r)t] - 1\}\right) \times$$

$$\left(p_0 + \frac{b + a}{(g - r)^2}\{\exp[(g - r)t] - 1\}\{\exp(\exp[(g - r)t] - 1) - 1\}\right)$$

for new investment is when the increase in equities affects the level of interest.

(iii) The market for bonds. The fundamental difference between equities and bonds is that the future cash flow from bonds is determined by a financial contract and is independent of the future rate of growth as long as a decreasing rate of growth does not threaten the financial ability of the debtor, which then implies some risk to the bond. Normally, however, bonds are less risky than equities. Without going into mathematical details, it is easy to see that, if the expected growth rate diminishes and the current cash flow is unaffected by this, we would expect an increase in demand for bonds, thereby diminishing the market rate of interest. As we have mentioned earlier, this will have a stabilizing effect on the demand for new capital and the market for equities. The stability will not be total since the equity owners will face a loss in value when restructuring their portfolios.

To accept this argument, however, we must disregard the fact that deteriorating expectations will normally indicate decreasing liquidity in the future but will in general also be based on decreasing current flow of liquidity. The first property will usually increase the subjective uncertainty of the future and the second property will imply less inflow of liquid resources to the financial market. If the latter problem exists we would expect increasing market rates of interest and thereby a worsening of the effects on the markets for new capital and equities. In this case the rate of interest is the price for diminishing liquidity.

In our consideration of production conditions we showed how decreasing capacity utilization would lead to a weaker effect on labour demand. This implies decreasing profit ratio and increased need of liquid resources relative to the turnover. There will also be a decreasing demand for labour which will probably result in an increase in relative saving and a greater tendency to hold savings in liquid forms. It is therefore reasonable to believe that the agents will act in favour of increasing rates of interest rather than in the opposite way when expectations of the future growth deteriorate. However, when expectations are improving, we would expect agents to act so as to reinforce the upward movement.

(iv) Gross substitution in financial markets. The stability condition for current prices proposed by Metzler (1945) and extended to include expected prices by Arrow and Nerlove (1958) states that goods must be gross substitutes in current as well as future prices. This condition is normally adopted by financial theorists. As we have discussed earlier, Tobin (1982) used the condition to guarantee a counteracting force when expectations of profit deteriorate, thereby reducing the investment demand.

However, there are some critics of this approach. Blanchard and Plantes (1977) argued from an econometric point of view that financial assets are never substitutes. Samuelson and Sato (1984) discuss the problem within the lines of hedonistic prices. From a theoretical point of view the latter approach is very interesting. The standard problem of hedonistic prices is that consumers judge the quality of different goods with the help of their respective prices, and the prices then enter the utility function. We shall then face the following maximization problem if we include the demand for cash balances:

$$\max_{M,x} U(M, x_1, \ldots, x_j p_1, \ldots, p_n) \tag{4.24}$$

subject to

$$rM + \sum_1^n p_j x_j = Y$$

Samuelson and Sato show that when prices enter the utility function as positve arguments, the integrability conditions are unattainable.

The portfolio choice is even worse. Wealth is determined by the asset prices: increasing asset prices imply increasing wealth. At the same time asset prices must form the left-hand side of the budget restriction. However, if we disregard the current flow of income, wealth will also appear on the right-hand side of the budget restriction. Thus the assumption of gross substitution *a priori* seems rather problematic.

Let us look at the three kinds of assets discussed in this chapter: new investment goods, equities on the second-hand market and bonds on the second-hand market. Stability implies that a change in the expected growth of demand which, *ceteris paribus*, will change the volume of investment is counteracted by price changes in equities and bonds. The alternative cost of the respective assets is the rate of return on the other assets.

If expectations of demand are improving this will have an effect on new investments and equities equivalent to a decrease in price. Given the market price of bonds, the capital value of investment goods will increase and hence investment will increase. If, instead, the new internal rate of return on capital was used as a discount rate in the bond market, the bond price will fall and the interest rate will rise until it reaches the new internal rate of return on capital. First, this will cause losses to bond-holders. Second, the only way of realizing this higher rate of return is to realize the investment. This is possible because the prices of investment goods are not determined in the financial market. The only way that the bond market can counteract the investment demand is if the price of bonds decreases by more than is necessary to keep the rate of return on bonds at the same level as the internal rate of return on capital.

The same arguments hold in the opposite case when expectations deteriorate. This is equivalent to a price increase on investment goods, but the decrease in investment can be counteracted if the excess supply in the market for investment goods makes the prices fall. However, it seems reasonable to believe that prices will follow the same pattern as discusssed in Chapter 2 for capital-intensive production. In this case the prices of equities will fall until the rate of return is the same as for bonds. Alternatively, bond prices will rise and the rate of return on bonds will decrease.

There is no way in which a financial market where the agents are well informed and rational in the usual neoclassical sense could play a stabilizing role with respect to changes in expectations. The expectations will form the new equilibrium prices which will imply a changed volume of investments. According to our discussion of short-run production functions in Chapter 2, where we showed the existence of stable unemployment equilibria and also showed that a decrease in demand relative to expectations implies a diminished ratio of profits to total value added, expectations based on *ex ante–ex post* comparisons may aggravate a deteriorating process without any counteracting forces from the financial market in the absence of a stabilization policy imposed by the government.

(b) *Increasing risk of capital*

When technical growth is high or accelerating the growth of demand must be equally high. This is the result of Harrod's theory of warranted rate of growth. In Chapter 2, Tables 2.1 and 2.2, we presented the results of some empirical tests based on a Cobb-Douglas production function with a Hicks-neutral technical change. It is interesting to see that the later period, 1972–84, displays a lower technical growth than the whole period 1961–84 — in fact it is negative. When we investigate 8 year periods in Table 2.2 we can see a systematic decrease in the growth factor except for the last period. The only time when we can interpret the growth factor as 'technical growth' is when we are following an equilibrium path, i.e. that at each instant *ex post* production equals *ex ante* planned production. Otherwise the growth factor will measure the growth of demand.

When we discussed the long-run development of the short-run production function, we found it plausible to assume that the 'fixed' volume of labour increased relative to the total volume of employment (see Fig. 2.11). The result is then a higher elasticity of labour productivity with respect to short-run variations in production. This in turn results in greater variations in operating surplus and profit share. Technical growth then implies that the risk to production capital increases.

At the end of Chapter 2 we introduced a production sector in which we assumed that production was proportional to employment and there was no technical growth. Most industrial countries have experienced a substitutional drop in industrial employment during the 1970s and 1980s. One explanation of this could be that there has been a structural change in demand in favour of goods from the low-productivity sector. To some extent this is true, but if it was the complete truth there would have been no unemployment. We believe that the reduction in employment in the industrial sector reflects the increased risk to production capital.

To meet this risk increase in the short run firms may hold a larger volume of financial assets, which then become complementary to production capital. In the long run they try to reduce competition by mergers. Both these strategies will decrease the growth of capital stock on the aggregate level, but firms adopting the latter strategy will endogenize a greater part of the risk increase inherent in technical growth. Thus, instead of Fig. 2.1 we have the situation shown in Fig. 4.10. The optimum labour productivity grows but the elasticity of labour productivity with respect to production also increases and thereby creates a much larger variation in the profit share. The increased risk to production capital then leads to a lower rate of growth of capital stock which in turn may lead to higher inflation rates in both the Verdoorn interval to the left of L^* and the Phillips interval to the right of L^*.

4.2.3 The financing decision in non-financial enterprises

The policy for managing capital is usually divided into investment policy and financing policy. Following the columns of the non-financial enterprises in the B and F matrices we have real and financial investments. The relative magnitudes of the entries are dependent on the portfolio decision. As mentioned above the rows for non-financial enterprises form the budget restriction on portfolio selection. To simplify notation we can express the budget identity at time t as

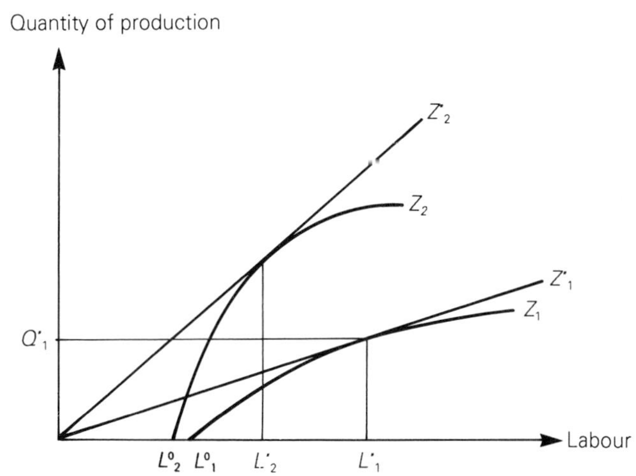

Fig. 4.10 Effect of labour-saving technical change when demand for output increases less then the productivity.

$$I + \frac{dB^{d}}{i_{d}} + \frac{dB^{f}}{i^{f}} + dm + rE = \Pi + dD + dE + i(B - D) \quad (4.25)$$

where I is net investment, dB/i is the change in financial securities (superscripts d and f denote domestic and foreign respectivly), dm is the change in cash balance, iD is the interest payment on financial debt, which can be separated into long and short-term debt, rE is the dividend to shareholders and Π is the net operating surplus.

The fundamental factor in the budget identity of non-financial enterprises is the net surplus, or gross surplus if we consider gross investments. This factor is also the most important in the portfolio selection of the company. As we found in the analysis of production the net surplus *ex post* is a residual. *Ex ante* we expect a yield which is maximal in relation to our investments, but *ex post* it will be a residual provided that we are not in equilibrium. Therefore a discrepancy between the firm's *ex ante* and *ex post* surpluses will not only induce revisions of the portfolio, but will also influence the current liquidity position and therefore the financial conditions for investment. This aspect is still more important when the 'solidity' i.e. the ratio of equity to total debt of the firm is poor. Services payment on financial debt is contracted in nominal terms while the income on working capital is determined by the market.

With regard to the financing of investments by share issues we take the usual view that this is an indirect method of removing short-term debt. In this case there could be a change in the market valuation of the company and therefore a change in the conditions for long-term finance relative to those imposed when the investment expenditure was undertaken. Dividends to shareholders are not normally based on the market valuation of the equities. This would be relevant if the market valuation were based on uniform information regarding the long-term prospects of the firm in question. The dividends are in general completed by a bonus issue to the current owners.[8]

We shall discuss the financial markets in more detail in Chapter 5, but we can conclude here, following Modigiani and Miller (1958), that the market valuation of equities is in principle independent of the strucutre of finance as long as the *ex ante* and *ex post* operating surpluses are equal. Given uniform information and risk distribution there will be an optimal relation between the risk to borrowers and the risk to shareholders. Given the yield of total capital, interest rate on borrowed capital and interest claims on equity, we obtain the optimum *ex ante* rate of debt from the following formula:

[8] Werin (1985) maintains that dividends are based on the market valuation. In April 1990 the average direct yield on equities on the Swedish stock exchange was about 3.5%.

$$\frac{D}{E} = \frac{r^{e} - r^{t}}{r^{t} - r^{d}} \tag{4.26}$$

where D is debt, E is equity, $D + E$ is total capital, r^{e} is interest claims on equity, r^{t} is expected rate of return on total capital and r^{d} is interest on debt. Differentiation with respect to the three different interest rates gives

$$\frac{d(D/E)}{dr^{e}} = \frac{1}{(r^{t} - r^{d})^2} \tag{4.27a}$$

$$\frac{d(D/E)}{dr^{t}} = \frac{1 - (r^{t} - r^{e})}{(r^{t} - r^{d})^2} \tag{4.27b}$$

$$\frac{d(D/E)}{dr^{d}} = \frac{1}{(r^{t} - r^{d})^2} \tag{4.27c}$$

The three derivatives are all positive. If we differentiate with respect to the difference between interest claims on equity and the expected rate of return on total capital and the difference between the rate of return of total capital and the interest on borrowed capital we have

$$\frac{d(D/E)}{d(r^{e} - r^{t})} = \frac{1}{(r^{t} - r^{d})^2} \tag{4.28a}$$

$$\frac{d(D/E)}{d(r^{t} - r^{d})} = \frac{-(r^{e} - r^{t})}{(r^{t} - r^{d})^2} \tag{4.28b}$$

An increase in the claims of shareholders relative to the expected rate of return on total capital will increase with the ratio of debt to equity. Expression (4.28a) shows that all three rates of return have a positive effect on the debt-to-equity ratio.

The above calculations indicate that when the economy experiences stable growth there will in general be an increase in the debt-to-equity ratio. However, we can expect the ratio to decrease when uncertainty of the future increases. If the ratio is high and the *ex post* result is worse than the *ex ante* expectation, the liquidity position will deteriorate faster than if the ratio had been lower. This will in turn reduce the planning horizon of the firm even if it succeeds in surviving. This point is clarified in Figs 4.11 and 4.12.

Figure 4.11 illustrates the relationship between the ratio of external debt to equity and interest claim on equity. In Fig. 4.12 the ratio of debt to equity is related to the interest rate of external debt. We can say that Fig. 4.11 reflects the dependence of the borrower's risk on the debt-to-equity ratio, while Fig. 4.12 shows the dependence of the lender's risk.

These two figures show that both equity owners and lenders will face an increased liquidity risk when the debt-to-equity ratio increases. We assume the following relationships:

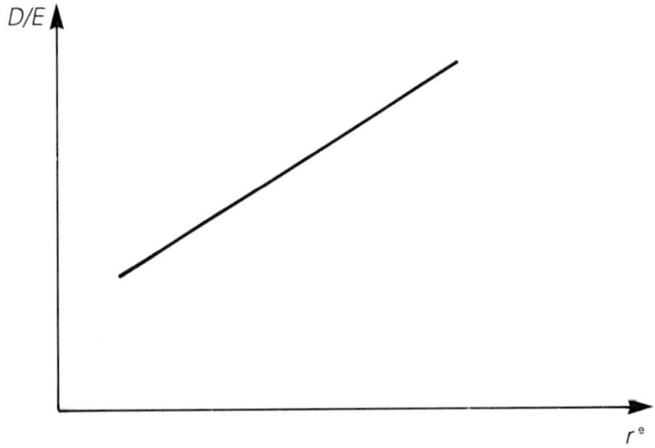

Fig. 4.11 Debt to equity ratio as a function of the interest claims on equities. Assumptions: r^t and r^d are given.

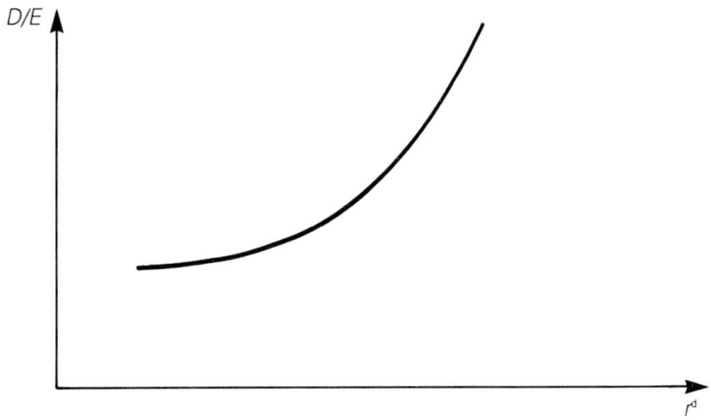

Fig. 4.12 Debt to equity ratio as a function of interest on debt. Assumptions: r^t and r^e are given.

$$r^e = f\left(\frac{D}{E}\right) \quad f' > 0$$
$$r^d = g\left(\frac{D}{E}\right) \quad g' > 0 \tag{4.29}$$

We then obtain the following differential equation:

$$\frac{D'}{E} = \frac{f(D/E) - r^t}{r^t - g(D/E)} \tag{4.30}$$

Simplifying notation and linearizing around equilibrium we can write

$$\mathring{d} = \frac{\alpha d}{\beta d} \tag{4.31}$$

provided that $f(D/E) \geqslant r^{\mathrm{t}}$ and $g(D/E) \geqslant r^{\mathrm{t}}$. The solution

$$d = k_0 \exp\left(\frac{\alpha}{\beta} t\right) \tag{4.32}$$

shows that stability is attained when $\alpha = 0$, i.e. when claims on equities are equal to the rate of return on total capital. From Figs 4.11 and 4.12 it is easy to see that the function is more sensitive to changes in the rate of interest on external debt than to changes in claims on equity. However, the central issue is that both rates work in the same direction to increase the instability of the ratio of external debt to equity. However, if the interest rate on external debt is greater than the rate of return on total capital the ratio of debt to equity will decrease.

Let us adopt another approach. Assume that claims on equity and interest on external debt are both affected by expectations of the rate of return on total capital, and that these expectations are positively related to the current rate of return, i.e.

$$r^{\mathrm{e}} = h_1(r^{\mathrm{t}}) \qquad h'_1 < 0 \quad h_1(r^{\mathrm{t}}) \geqslant r^{\mathrm{t}} \tag{4.33a}$$

$$r^{\mathrm{d}} = h_2(r^{\mathrm{t}}) \qquad h'_2 > 0 \quad h_2(r^{\mathrm{t}}) \leqslant r^{\mathrm{t}} \tag{4.33b}$$

It is reasonable to assume that claims on equity increase when the agent experiences increased uncertainty about the future. It is also reasonable to assume that lender's risk, and therefore the interest rate on external debt, increases when expectations of future yield on total capital deteriorate. Inserting equation (4.32) into (4.25) and differentiating with respect to r^{t} gives

$$\frac{\mathrm{d}(D/E)}{\mathrm{d}r^{\mathrm{t}}} = \frac{(h'_1 - 1)(r^{\mathrm{t}} - h_2) - (1 - h'_2)(h_1 - r^{\mathrm{t}})}{[r^{\mathrm{t}} - h_2(r^{\mathrm{t}})]^2} \tag{4.34}$$

Given the assumptions made in Equation (4.32) the derivative will have a negative sign provided that $h_2(r^{\mathrm{t}}) < r^{\mathrm{t}}$. In the opposite case the sign of the derivative will depend on the relative sensitivity of equity owners and lenders with respect to changes in the yield of total capital. If the sensitivity of equity owners is greater than that of lenders the derivative will have a positive sign.

The conclusion of the above discussion is that if firms make their financing decisions in the way proposed in expression (4.25) they will generally increase their future liquidity restrictions when future expectations deteriorate and if the lenders react in such a way that a decrease in solidity (or ratio of equity to total debt) will increase the lender's risk. This will aggravate the process of increasing debt relative to equity.

We could also use expression (4.25) to study the behaviour of equity claims with respect to different interest rates and debt-to-equity ratios. We rewrite the expression

$$r^{e} = r^{t} + (r^{t} - r^{d})\frac{D}{E} \tag{4.35}$$

Let us use this expression to study the effect of changes in the rate of return on total capital on the claims on equity. A simple differentiation shows that an increase in the yield of total capital will, *ceteris paribus*, increase the claims on equity. Furthermore, the claims on equity will always decrease when interest rate on external debt increases:

$$\frac{\delta r^{e}}{\delta r^{t}} = 1 + \frac{D}{E} \tag{4.36a}$$

$$\frac{\delta r^{c}}{\delta r^{d}} = - \frac{D}{E} \tag{3.46b}$$

Let us assume that the interest rate on external debt is a function of the yield of total capital, i.e. the interest rate is related to the lender's risk. We can write

$$r^{d} = h(r^{t}) \qquad h' < 0 \tag{4.37}$$

Inserting (4.37) in (4.11) and differentiating with respect to r^{t} we arrive at

$$\frac{\mathrm{d}r^{e}}{\mathrm{d}r^{t}} = 1 + \frac{D}{E}(1 - h') \tag{4.38}$$

The equation implies that the effect of an increase in the yield of total wealth on claims on equity will be greater when the debt-to-equity ratio increases.

The last point is important according to the Minsky theory of increasing vulnerability in periods with 'sustained growth' (Minsky, 1964). In a period of sustained growth, when profit shares decrease and there is high activity of investment with increasing debt-to-equity ratio, a sudden change in the yield of total capital will have an aggravated effect through relation (4.37), particularly if the firm's policy has been use to increasing debt-to-equity ratio as a means of increasing the yield on equities according to (4.34). We are reminded of the inverse relationship of claims and equities relative to equity prices.

Summarizing the discussion of the firm's financing decisions we find that, given the expectations of the future production performance, the firm may choose the debt-to-equity ratio to affect the yield on equities. An optimal choice at one instant or sequence of instants may increase the future instability of equity prices and aggravate the future liquidity position. The reason for this, of course, is that the yield of real capital, i.e. total capital if we disregard financial capital acquired by the firm, is market determined at every instant while the debt is a financial contract with interest at best independent of the yield of total capital (sinking fund finance) or inversely related to the rate of return on real capital (floating fund finance) (Keynes, 1936, Chapter 8; Vickers, 1974).

REFERENCES

Ando, A. and Modigliani, F. (1963) The life cycle hypothesis of saving: aggregate implications and tests. *American Economic Review*, **53**, 58.

Arrow, K. J. and Nerlove, M. (1958) A note on expectations and stability. *Econometrica*, **26**.

Blanchard, O. J. and Plantes, M. K. (1977) A note on gross substitutability of financial assets. *Econometrica*, **45**.

Dahlman, C. J. and Klevmarken, A. (1971) *Den Privata Konsumtionen 1931–1975*, Almquist and Wicksell Uppsala.

Debreu, G. (1974) Excess demand functions. *Journal of Mathematical Economics*, **1** (1).

Dusenberry, J. S. (1952) *Income, Saving and the Theory of Consumer Behavior*, Harvard Ec. Stu., Cambridge, MA.

Ekstedt, H. and Westberg, L. (1988a) Interaction between economic growth and financial flows: presentation of a model analysing the impact of short-term financial disturbances on economic growth, in *Economic Modelling in OECD Countries* (ed. H. Motamen), p. 229, Chapman and Hall, London.

Ekstedt, H. and Westberg, L. (1988b) A dynamic model for the interrelations of real and financial growth. Paper presented at the International Conference on Economic Modelling, London, 28–30 March, p. 24 (in press).

Enthoven, A. C. (1956) Monetary disequilibrium and the dynamics of inflation. *Economic Journal*, **66**.

Enthoven, A. C. and Arrow, K. J. (1956) A theorem on expectations and the stability of equilibrium. *Econometrica*, **24**.

Fama, E. (1970) Multiperiod consumption–investment decisions. *American Economic Review*. **60**.

Goldsmith, R. (1955) *A Study of Savings in the United States*, Princeton University Press, Princeton.

Grimmet, G. R. and Stirzaker, D. R. (1982) *Probability and Random Processes*, Clarendon Press, Oxford.

Hicks, J. R. (1939) *Value and Capital*, Clarendon Press, Oxford.

Kaldor, N. (1940) A model for the trade cycle. *Economic Journal*, March.

Keynes, J. M. (1936) *General Theory of Employment, Interest and Money*, Cambridge University Press, Cambridge.

Kuznets, S. (1942) *Uses of National Income in Peace and War*, National Bureau of Economic Research, New York.

Kuznets, S. (1946) *National Product since 1869*, National Bureau of Economic Research, New York.

Lavacic, R. and Rebmann, A. (1982) *Macroeconomics*, Macmillan, London.

Markowitz, H. (1959) *Portfolio Selection: Efficient Diversification of Investment*, Wiley, New York.

Mas-Colell, A. (1985) *The Theory of General Economic Equilibrium: A Differentiable Approach*, Cambridge University Press, Cambridge.

Merton, R. C. (1973) An intertemporal capital asset pricing model. *Econometrica*, **41**.

Merton, R. C. (1982) Microeconomic theory of investment, in *Handbook of Mathematical Economics* (eds. K. J. Arrow and M. D. Intriligator), North Holland, Amsterdam, Chapter 13.

Metzler, L. A. (1945) Stability of multiple markets: The Hicks' conditions. *Econometrica*, **13** (4).

Minsky, H. (1964) Financial crises, financial systems and the performance of the economy, in *Private Capital Markets* (eds. I. Friend, H. Minsky and V. Andrews), Prentice Hall, Englewood Cliffs, NJ.

Modigliani, F. and Miller, M. H. (1958) The cost of capital, corporation finance and the theory of investment. *American Economic Review*, 48.

Morishima, M. (1973) Consumer behavior and liquidity preference in theory of demand, in *Theory of Demand: Real and Monetary* (eds. M. Morishima *et al.*), Clarendon Press, Oxford.

OECD (1988) *Economic Outlook*, **44**, 168.

Pollak, R. A. (1970) Habit formation and dynamic demand functions. *Review of Economic Studies*, **38**.

Samuelson, P. A. and Sato, R. (1984) Unattainability of integrability and definiteness conditions in the general case of demand for money and bonds. *American Economic Review*, **74**.

Sanchez, D. A. (1968) *Ordinary Differential Equations and Stability Theory*, W.H. Freeman, San Francisco, CA.

SIND PM (1986) *Investeringar, Finansiering och Lönsamhet i 22 Företag (Investment, Financing and Profitability in 22 Companies)*, Statens Industriverk, Stockholm.

Smithies, A. (1945) Forecasting postwar demand I. *Econometrica*, January.

Tobin, J. (1980) *Asset Accumulation and Economic Activity*, Basil Blackwell, Oxford.

Tobin, J. (1982) *Money and Finance in the Macro-economic Process*, Nobel Foundation, Stockholm.

Vickers, D. (1974) Disequilibrium structures and financing decision of the firm. *Journal of Business Accounting*, **3**.

Werin, L. (1985) Perspektiv på budgetunderskottet. Del. 3, DsFi 1953: 29.

Wicksell, K. (1916) *Föreläsningar i Nationalekonomi* Gleerups, København. 5th edn. (*Lectures in Economics* English translation 1936, London School of Economics, London.

5

Financial flows and stocks

5.1 FINANCE IN THE TRANSACTIONS TABLE

We have seen how cyclical variations in income growth and structural changes in the production system in the medium and long term lead to cyclical and medium-term variations in the growth rate of the factor incomes in the various institutional sectors. These changes also produce both cyclical and more durable changes in the growth rates for expenditure on final goods.

The propensity for expenditure on final goods in the various institutional sectors is influenced by the net income transferred from other sectors and the possibility of increasing the current liquid resources by credit from other institutional sectors. Expenditure on final goods is also influenced by the sectors' possibilities of undertaking expenditures on financial assets. These financial transactions are registered in the finance matrix in the transactions table.

The very simplified transactions table in Fig. 5.1 can show how the household sector can enlarge its current liquid resources R_H in the form of wages $W_N + W_F$ with credits F_{HF} from financial institutes. The household sector may also want to increase its financial assets by increased deposits F_{FH} in financial institutes and increased claims F_{HN} on non-financial institutes. The household sector uses the rest of its disposable income for consumption expenditure C. Current liquid financial resources in non-financial enterprises R_N consist of the sector's current surplus π_N, loans from financial institutes F_{NF} and loans from households F_{NH}. These resources are used for gross investment in real capital I and in financial assets, in the form of claims F_{FN} on financial institutes. The current financial resources of the financial enterprises consist of current surplus and credits from non-financial institutes and households. These resources are used for credits to households and non-financial enterprises. The input–output account for the financial enterprises is not complete in the example as no output balancing the input due to factor incomes $W_F + \pi_F$ is recorded. The outputs are

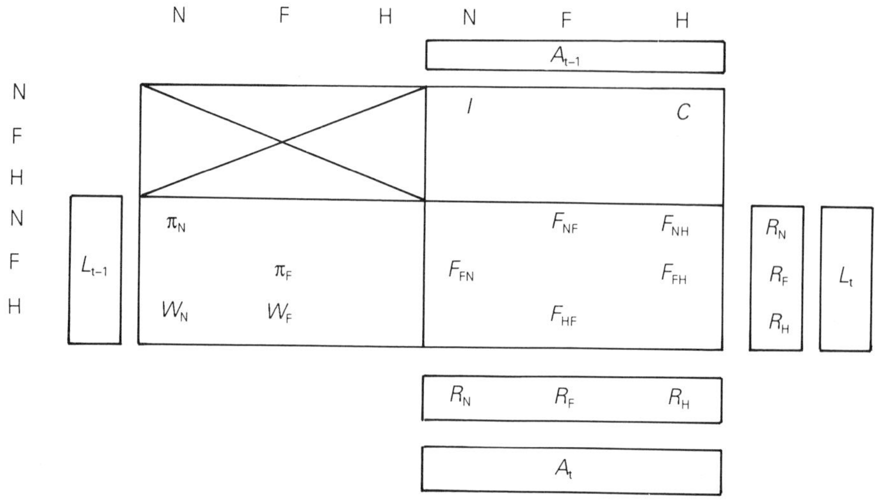

Fig. 5.1 A simplified transactions table: N, non-financial enterprises; F, financial enterprises; H, households; *I*, investment in real capital; *C*, consumption; π_N, profits in non-financial enterprises; π_F, profits in financial enterprises; W_N, wages earned in non-financial enterprises; W_F, wages earned in financial enterprises; F_{NF}, credits from financial enterprises to non-financial enterprises; F_{NH}, credits from households to non-financial enterprises; F_{FN}, credits from non-financial enterprises to financial enterprises; F_{FH}, credits from households to financial enterprises; F_{HF}, credits from financial enterprises to households; R_N, current financial resources in non-financial enterprises; R_F, current financial resources in financial enterprises; R_H, current financial resources in households; A_{t-1}, assets accounted for at the end of period $t-1$; A_t, assets planned for the end of period t; L_{t-1}, liabilities accounted for at the end of period $t-1$; L_t, liabilities planned for the end of period t.

assumed to be intermediary services in the *A* matrix, but no transactions are recorded in this matrix.

Although the transactions table in Fig. 5.1 is extremely simplified, it illustrates how cyclical changes of income distribution betweeen current surplus in the enterprises result in cyclical variations in current financial resources in the case when factor income variations are not balanced by changes in credit. It also shows how current liquid resources are used for expenditure on final goods and financial investments, except in financial enterprises where no expenditures on final goods are registered.

Although the example in Fig. 5.1 refers to a very simplified model of an economy the stock of assets and debts is included. This has been done because the agents' choice of financial investments and expenditure on final goods may not be influenced by the flow of current liquid financial resources. Instead, current financial resources may influence agents' expectations of future resources, which in turn will influence the valuation of assets. It may be these asset prices which are the main basis for determining planned expenditure on finished goods.

Figure 5.2 shows how the expectations of future current liquid resources F^{R}_{t+1} in the matrix are based on the development of liquid resources during earlier periods F_{t_0}, F_{t-1}, F_{t-2} and so on, and how these expected values influence the current valuations of assets A_t (for simplicity, the liabilities have not been included in the figure).

(a) *Sources of funds*

The simplified model of the economy shown in Fig. 5.1 has three sources of funds: funds built up from the sectors' own savings, and external finance in the form of either issues of bonds and equities or loans in financial institutes. The first of these types of external finance has been called 'primary securities' and the second 'indirect securities' (Gurley and Shaw, 1960).

The source of funds analysis can show that, in addition to its importance for the yield on capital, the operating surplus of the enterprise is decisive for business investments because of its effects on the equity capital of enterprises. The equity capital can be increased by business saving, which is the funded part of the operating surplus, or by new capital issues. The alternative way of financing investments is to increase debts. However, the possibility of financing an investment expansion by increasing the relation between debt and equity is dependent on the current economic situation. If the production growth rate is high and stable, an increase in the debt-to-equity capital ratio would not cause any problems. The yield on business total capital would be high and greater than the interest rate for loans. In

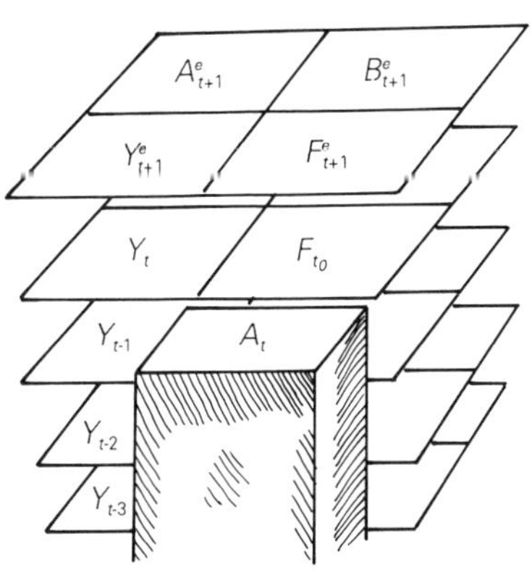

Fig. 5.2 The transactions table in the time dimension: A^{D}_{t+1}, expected transaction of intermediary goods in period $t+1$; B^{e}_{t+1}, expected transaction of final goods in period $t+1$; Y^{e}_{t+1}, expected transaction of factor incomes in period $t+1$; F^{e}_{t+1}, expected financial transaction in period $t+1$.

such a situation a high debt-to-equity capital ratio will mean a high yield rate on business equity capital. This will be a result of the so-called leverage effect. However, if the economic situation is characterized by a slow growth rate and uncertainty about future development, the yield on total business capital may be lower than the interest rate on loans. Therefore the yield on business equity capital will be lower the higher is the debt-to-equity ratio. The reason for this is that if the enterprise becomes insolvent the creditors have a right to their loans with interest, and it is only the remainder of the capital that is available for the owners.

There may be a particular risk to the ability to maintain economic expansion during a long growth process if the business sector has financed capital building with increasing debt ratios. If there is a hidden weakness in the growth process, perhaps because of bottlenecks in the supply of raw materials, energy or labour, the expectations of future income and profit growth may deteriorate, which will result in falling prices on equities and bonds.

In such an economic development therefore, can a decreasing profit share not be substituted by an increasing debt ratio? The only possibility of financing a continued growth in investment would be an increase in new equity capital issue. In order to stimulate investment during the period of low growth rate at the end of the 1970s and the beginning of the 1980s, there was a reform in the tax laws in order to stimulate the demand for equity in some industrial countries. However, it should be noted that during a recession there is probably a very low demand for new capital in the industry. Deficits in the balance of payments and in the national budget may lead to high interest rates and make the business sector more interested in financial investments than in investments in real capital.

(b) *Short and long borrowing*

We have seen that there is a complementary relation between the sources of funds in the form of internal business saving, equity financing and financing by credits from other institutional sectors. The necessity of stable relations between these sources of finance has been stressed by the propensity to keep a satisfactory relation between the enterprises' own capital and the total capital (the sum of debts and equity capital), the so-called 'solidity ratio'. However, we have noted that this ratio may fluctuate according to the growth rate in the economy. In times of expansion it may decrease without creating problems for the enterprises. In times of recession a low value of the solidity ratio may lead to financial difficulties and in extreme cases to insolvency.

The complementary relations between different sources of funds on the credit market have also been stimulated by customs concerning the division of labour between different types of credit institutes. Different types of investment are generally financed in special types of institutes. The short credits necessary for business working capital are usually financed in commercial banks, even if the enterprises have a high degree of self-finance. Investment in buildings may be financed in commercial banks during the construction period. After the buildings are complete the bank loans may be refinanced in other types of credit institutes at lower costs as mortgages in the new buildings will give good security. These types of credit institutes generally specialize in longer types of credits than has been customary in commercial banks. They also claim to be secure. They have often been characterized as 'capital market institutes', as their most important financing has been in bond and equity markets and the market for long credits to the building industry.

Some of these types of institutes specialized in financial services which were of great importance in the early phase of industrialization. Some were of a regional type such as savings banks which were important in many countries for both credits in agricultural regions and credits for house building in urban regions. Modern industrial development has led to a strong expansion in insurance and pension institutes. The growth in the development of credit institutes outside banks was particularly stimulated by the rapid economic growth after the Second World War during the 1950s and 1960s. Its importance for economic development and for the possibilities of influencing this development with the traditional methods of monetary policy was a central problem discussed in, for example, the Radcliffe Report in the United Kingdom in 1959.

The low growth ratio in many industrial countries during the 1970s and 1980s has led to a need for consolidation and concentration among institutes of this type. In some cases public pension systems and pension funds have hastened this development. The strong expansion and internationalization of banks that was stimulated by the oil price crises during the 1970s and the accompanying large increase in international credits—a tendency which has continued to increase during the 1980s—has also influenced this development. In addition to the internationalization of the credit system there is also a tendency to concentration in large units where traditional banking services are combined with services which traditionally have been the province of other types of institutes such as insurance companies, savings banks etc.

(c) *Cash flow and investment finance*

The type of credit that the borrower will demand depends on the expectation of the time structure of future cash inflow. If he can arrange his transactions in such a way that he will receive cash revenue before he will have to make interest and mortgage payments for his loans, his liquidity position will not be in danger. When the expected cash inflow exceeds the interest contractually due and the mortgage payments in each period, the firm uses what is known as 'sinking fund finance'. The firm's good liquidity position which will follow from sinking fund finance will influence the present value of the assets of the firm above its supply price at any given rate of time discount.

It is quite possible that even if the debts are in sinking fund finance contracts, so that the expected cash inflow exceeds the obligations in each period, the borrowers may wish to keep additional liquid assets as a precautionary measure. The reason for this may be a precautionary motive for holding extra liquidity. It is possible that expected cash inflow is not sufficient to meet the contractually due interest and mortgage payments in each period. Such debt contracted payment obligations are called 'floating fund finance'. Such a situation forces the borrowing sector to seek new credit in order to refinance its position. The policy could be rational if it is possible to obtain short credit cheaper than long credit. The reason could also be that there is insufficient credit capacity on the market for long credits or that the lending preferences of the institutions on the market for long credit are not suitable for the sector demanding credit. An example of such a situation was the difficulty that the Swedish government encountered in the 1950s in borrowing on the long market before the creation of the Allmänna Pensionsfonden (General Pension Fund) increased the lending capacity on the long market in Sweden.

It has been pointed out that a stable financial development may deteriorate when business sources of finance are dominated by sinking funds finance. If the growth rate goes down sinking funds finance may become floating funds finance (Davidson, 1978, p. 415). If sales expectations are not met the cash inflow may be insufficient to meet the obligations of interest and mortgage payments. New credits will be necessary and floating funds finance will develop. In this way disappointment in expected sales revenue not only affects current expectation and reduces the marginal efficiency of investment, but also creates liquidity problems. In such a situation business will be very vulnerable if the monetary policy is made more restrictive. Higher interest rates and unwillingness of the banks to supply the new credit necessary to refinance the firm's position may lead to insolvency.

It has been argued that such a weakening of the firm's financial position would characterize the end of a long growth process in which the need for intensively increasing credits has been hidden by rising bond and equity prices due to the stable growth rate and unchanged profit expectations (Minsky, 1964, p. 213). When a weakening in the growth process is reflected in an increasing deficit in the balance of payments or a higher inflation rate leads to a more restrictive monetary policy the financial position of industry is undermined. In such conditions the appearance of floating funds finance will be interpreted as the beginning of cyclical stagnation. The maintenance of sinking funds finance will secure bond finance and stable growth. (Davidson, 1978, p. 415).

It is interesting to compare Minsky's views with Keynes' critique of sinking funds finance:

> ... sinking funds, etc., are apt to withdraw spending power from the consumers long before the demand for expenditure on replacements (which such provisions are anticipating) comes into play; i.e. they diminish the current effective demand and only increase it in the year in which the replacement is actually made. (Keynes, 1936)

Therefore instead of being a condition for a stable growth of production, the maintenance of sinking funds finance will hold back the growth in demand. Perhaps Keynes' critique of sinking funds finance can be said to be a critique of a monetary policy which prevents a floating funds finance being maintained with easy money and low interest rates.

(d) *Uses of funds*

In the introduction to this chapter we noted how the uses of funds in different sectors consisted of their expenditures on final goods, transfers to other sectors and investments in financial assets. These different uses of funds can be identified in the matrices for final goods and financial transactions in the transactions table. Figure 5.3 shows four columns in the matrix for final goods and financial transactions.

Two forms of investment, gross investment in real capital in the matrix for finished goods and financial investment in the form of a change in bank deposits in the financial matrix, are registered the column for non-financial enterprises (NON). Just as it is possible to find complementary relations between different sources of funds, there are also complementary relations between the business sector's uses of funds. Such a relation may exist between increases in liquidity in the form of bank deposits and gross investment. There will probably be some time lag in this relation, such that investment in real capital is preceded by increases in liquidity (Kragh, 1967a, p. 36).

Transactions within the non-financial enterprise sector are not recorded in the financial matrix. Otherwise types of transactions could be found which could be of great importance for the development of an investment expansion. The regular credits between firms which are a result of the existence of several stages of production are transactions of this type. Firms may also make transactions on the share markets, which could be both complementary and substitutionary for investments in real capital. Share holdings in firms in branches with other types of cyclical demand which can lead to a more stable profit growth in both enterprises are examples of the first type. A substitutionary relation exists if the production capacity is

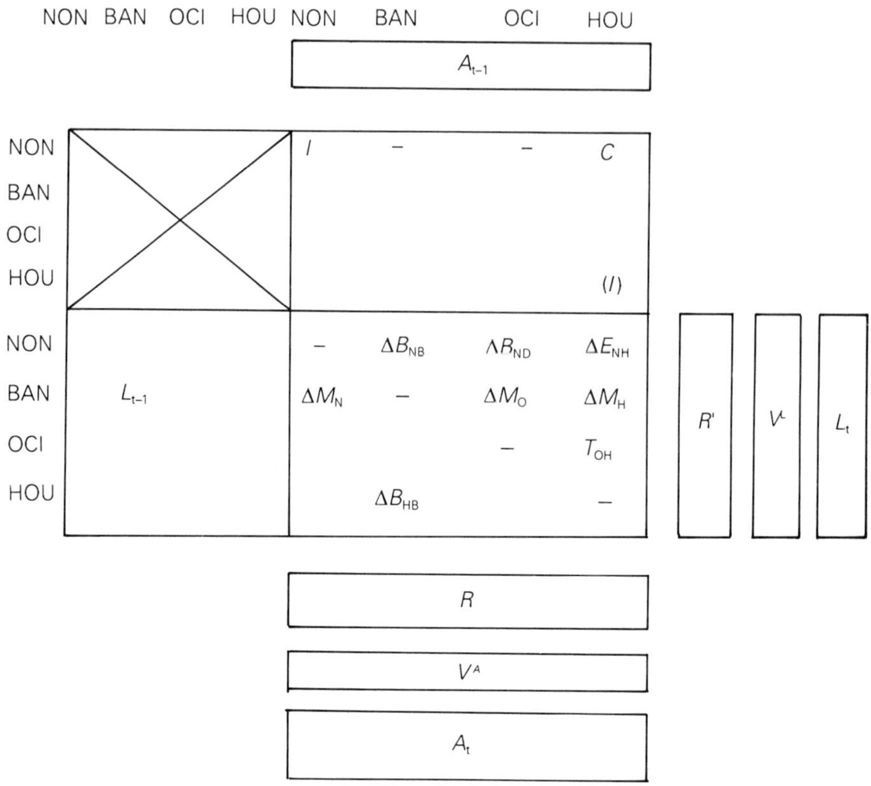

Fig. 5.3 A transactions table for uses of funds: ΔM_N, change in bank deposits in non-financial enterprises; ΔM_D, change in bank deposits in other credit institutes; ΔM_H, change in bank deposits in households; ΔB_{NB}, change in bank loans to non-financial enterprises; ΔB_{NO}, change in loans by other credit institutes to non-financial enterprises; ΔE_{NH}, change in households' holdings of equities supplied by non-financial enterprises; T_{OH}, households' fees to pension funds and other credit institutes; ΔB_{HB}, changes in bank loans to households.

increased by purchasing another company instead of increasing the real capital by investment.

Five items are registered in the household sector column (HOU) in the matrices for financial goods and financial transactions. Purchases of consumption goods and services from the sector of non-financial enterprises are included in the first row in the matrix for finished goods. The household sector's own investment in house construction has not been identified, as in the calculation of national accounts this item is often included together with other construction in the gross investment of non-financial enterprises. From a financial point of view it is of interest to note such internal investment in the household sector as it is one of the most important reasons for financial behaviour in the sector. The share transactions are recorded in the first row of the financial matrix and changes in its liquidity in the form of bank deposits are recorded in the second row (BAN). Transfers to other credit institutes are registered in the third row (OCI) as they are the current inflow in these institutes and the basis for its supply of long-term credit to non-financial enterprises.

The reason for separating the financial enterprises into two sectors—banks and other credit institutes—is that traditionally they have different roles in the credit market. The banks concentrate on risk credits which are formally short run while the other credit institutes act in the capital market and give long credit with formally higher securities. In all industrial countries the banks have the dominant role in supplying commercial credit while the capital market institutes dominate the bond markets. This sector includes different types of funds which have come to play a dominant role on the equity market.

Our analysis makes it possible to work with a highly aggregated sector distribution and to concentrate on a few strategic aspects of the portfolio behaviour of the public and the credit institutions. The banks' dominant assets are loans to the public, which are represented in the financial matrix by credits to non-financial enterprises and to households. Funds that are not demanded for such credits are invested in short bonds or are kept in deposits in the central bank if the banks are obliged to maintain a certain cash ratio. If the economic situation makes the banks restrict their credit to the public because the expectations of future demand have deteriorated, they may prefer to increase investment in short liquidity bonds. There may also be rules which require that banks hold government bonds or other types of bonds which have preference depending on the economic policy in their portfolios.

The banks' current financial resources are mainly transfer incomes and deposits from the public. The banks incur interest costs in order to obtain deposits from the public. The growth of profit in banks depends mainly on the differences between interest income and interest costs. In some industrial countries the banks' services to customers with respect to payments and portfolio administration are becoming even more important than their traditional interest income.

All types of financial enterprise which are not banks are brought together in the sector denoted other credit institutes. In most industrial countries it is dominated by capital market institutes like insurance companies and pension funds. Their transactions take place mostly on the bond market but also on the share market. In the model that we shall analyse, the simplified assumption has been made that the institutes invest primarily in long bonds which the non-financial institutes issue in order to consolidate the bank loans which finance that part of current real capital investment that is not financed by the non-financial enterprises' own saving. It is also assumed that the financial investments in the household sector are primarily on the equity market. These simplified assumptions have a realistic background in industrialized countries, where the capital market institutes are often developed in order to finance those sectors which can give good security for their loans by mortgaging real property or issuing bonds. Such sectors are, for example, the public sector and the house-building sector, industries with natural resources and energy production. In these countries most household financial saving comes from a small number of wealthy families, who are frequently owners of a major part of the industry. Their financial investments will naturally be restricted to equity markets.

We shall also assume that a large amount of household saving is channelled through insurance and pension funds to the capital market institutes in the sector for other credit institutes. The households' premium and fees payments depend on their income growth and it is the preferences of the capital market institutes which decide where and when these funds are invested in financial assets. These institutes are also assumed to keep liquid assets in the form of bank deposits. It is therefore possible to apply Keynesian liquidity theory to this transaction on the bond market. The same kind of theory can be applied to the household sector's choice between equities and money in the form of bank deposits if this is assumed to be partly governed by a speculative motive.

There are also assets in the form of bank deposits in the sector for non-financial enterprises. Although this sector may also have other types of financial assets, their cash liquidity should mainly be seen as a buffer in their transactions of goods and services and not for speculation on the financial markets. While a Keynesian liquidity theory is more appropriate in the sector for households and credit institutes, in countries with a less developed system of credit institutes it may be possible to apply some kind of quantity theory in the sector for non-financial enterprises. In an even more internationalized world such possibilities will be of minor importance.

5.2 THE CYCLICAL DEVELOPMENT OF FACTOR INCOME DISTRIBUTION AND ASSET PRICES

During the first phase of cyclical expansion profits in the sector for non-financial enterprises usually show a particularly high growth rate. There will be a strong displacement of the factor income distribution from wage earners to capital owners. Increasing dividends and income expectation will increase the demand on equity markets. As the need for new equity capital in this phase of the business cycle will be fairly low the result will be substantial increases in equity prices.

The capital market institutes, whose resource growth is mainly influenced by the growth rate of wages, may show a lower growth rate for their financial investments in this phase. However, they may have better investment opportunities in the bond market than in an overheated share market where there may be a slow growth rate of supply because of the high profit expansion and a low growth rate for investment in real capital if there is still a fairly low capacity utilization.

In later phases of the growth process the profit growth rate will decline and the profit share will decrease before the peak of the cycle. There will be an increase in the need for external finance in the business sector. As long as the growth rate is stable it will also be possible to obtain this finance without increases in the interest rates. The reason is that the increase in the growth of wage rate and ultimately in the share of wages in factor incomes will lead to increased financial resources in credit institutes, both banks and capital market institutes. It will not be difficult for business to obtain increasing bank loans and to consolidate these loans by bond flotation as demanded by capital market institutes. There is also an increasing supply of equity capital as the non-financial enterprises are eager to maintain a satisfactory ratio between their own capital and their debts. This increase in equity capital will be taken up in the household sector and in capital market institutes.

When the production growth rate decreases the increase in the wage share of factor income will be reinforced as the increase in wage rates continues because of the cyclical time lag in wage movements and the substantial reduction in profits. The latter influences profit expectations and leads to falling prices in security markets. These tendencies are strengthened if a general demand for increased liquidity develops. The possibility that the central banks will comply with the increased need for liquidity depends upon the development of the balance of payments in leading industrial countries, the exchange rates and the tendencies to increased inflation. The information flow will indicate variable tendencies in these factors and equity and bond prices will be extremely unstable.

In total, the decrease in the production growth rate which leads to a lower growth in financial resources as long as the financial policy is mainly passive will lead to a deterioration in income expectations on the one hand and to a major change in factor income distribution on the other hand. On the credit market the change in income distribution may lead to a transfer of financial resources away from investors active in more specialized types of assets to capital market institutions with broader portfolios including both equity and bond markets. The result will be that the change in the production growth rate will also lead to a change in the demand for bonds and equities and the sensitivity of this demand to the expectations of relative price changes. While the phases of a stable and acceptable growth rate are characterized by general expansion of demand equal for all types of assets, which leads to complementary relations, the stagnation phase leads to changes in income distribution and expectations which result in substitutions between different types of assets and a high sensitivity to expected price changes.

5.3 A KEYNESIAN MODEL FOR THE RELATION BETWEEN REAL AND FINANCIAL DEVELOPMENT

An analysis of the financial development resulting from changes in the production growth rate during the business cycle can be better understood using Kragh's model for the Keynesian liquidity function (Kragh, 1955, pp. 155–69). The model has three sectors: non-financial enterprises who make the real capital investment, households who represent saving and banks who are treated as credit intermediaries.

It is assumed that investment in period $t-1$ is financed by bank loans and that these bank loans are consolidated in period t by bond floatations sold at a price corresponding to the rate of interest in the preceding period. Therefore the current flow supply of bonds is equal to the investment in the preceding period. The demand for bonds is given by the liquidity preference function, i.e. a stock function. It is assumed that all current saving is invested in bonds. Therefore flow equilibrium in the bond market will depend upon how investment develops. If investment is equal in period $t-1$ and period t and there is equilibrium in the goods markets, so that current saving will be equal to current investment *ex ante*, supply and demand for bonds will be equal and the interest rate will be stable. If investment is changing there will be a disequilibrium in the bond market and the interest rate will change. Kragh analyses a situation when investment is increasing; there will be a surplus demand for bonds, as the supply equals the investment from the period $t-1$ and the demand is equal to investment in the period t, i.e. current saving.

The disequilibrium situation of the model example can be studied in the finance matrix of the transactions table shown in Fig. 5.4. Both variables that are included in the model and variables that do not enter the model are entered in the transactions table. The latter group are in parentheses. Investment in the period $t-1$ is assumed to be smaller than the planned

		NON	BAN	HOU
		I_t		(C_t)
NON			ΔL	ΔB
BAN		ΔM		
HOU	(Y)			

Fig. 5.4 Transactions table for a financial model economy *ex ante*.

investment for period t given in the transactions table. This means that the bank credits necessary for financing the investment in period t are greater than the credits given by the banks for financing the investment in the period $t - 1$. The increase in bank loans results in an equal increase *ex ante* among the assets in the NON sector. As the investment in period $t - 1$ which is to be consolidated by bond flotations during period t is less than the demand ΔB for bonds, which is registered *ex ante* in the household column in the finance matrix, there will be a demand surplus for bonds which will drive up their price. On that account the investors will be willing to sell bonds which will result in an increase of cash in the household sector. The column of the household sector in Fig. 5.5 registers this increase in cashholding, ΔM, and the increase in bondholding, ΔB, which *ex post* is equal to the supply of bonds planned in the NON-sector.

The result for the development of the interest rate can be studied using a liquidity preference curve (Fig. 5.6). ΔM shows the net increase in cash which is a result of the households' bond sales owing to their liquidity preference. The demand surplus creates bond price increases so that the market interest ratio is reduced from r_0 to r_1.

What would happen if there was a decrease in investments in non-financial enterprises? With equilibrium in the markets for goods and services, household savings *ex ante* would be smaller in period t than in period $t - 1$. Therefore the demand for bonds would be insufficient to cover the supply *ex ante* in enterprises. The result *ex post* would be that some of the bank loans received for financing investment in period $t - 1$ could not be consolidated in period t. The non-financial enterprises would have to pay back these loans by drawing on their bank deposits. This situation is illustrated in Fig. 5.7. The surplus of supply in the bond market will cause falling bond prices and an increase in the market interest rate, as can be shown by the liquidity preference curve (Fig. 5.8). Reductions of deposits in enterprises in order to reduce bank loans which cannot be consolidated by bond issues because of households' reduced resources to invest lead to increasing bond rates.

		NON	BAN	HOU
	X	I	–	(C_t)
NON	–		ΔL	ΔB
BAN	–			ΔM
HOU	(Y)			
			R	

Fig. 5.5 Transactions table for a financial model economy *ex post* with increasing money stock.

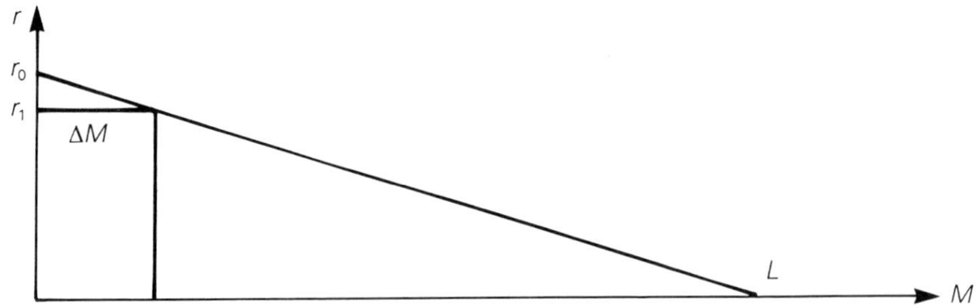

Fig. 5.6 Liquidity preference curve with increasing money stock.

		NON	BAN	HOU
		I	–	(C_t)
NON	–		$-\Delta L$	ΔB
BAN	–	$-\Delta M$		–
HOU	(Y)			–

Fig. 5.7 Transactions table for a financial economy *ex post* with decreasing money stock.

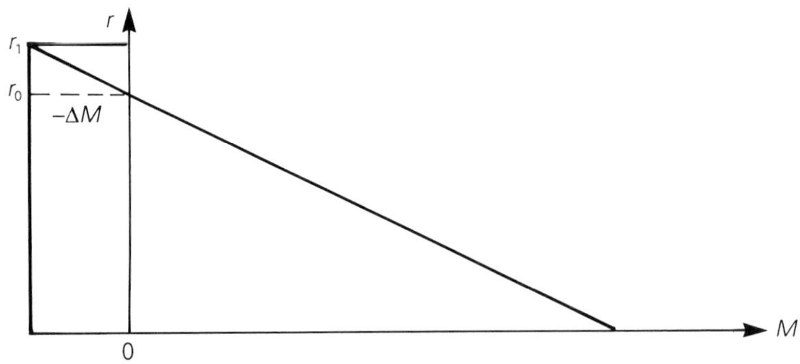

Fig. 5.8 Liquidity preference curve with decreasing money stock.

(a) *Short-run changes in the preferences for uses of funds and sources of funds*

During the business cycle the preferences for both financial investments and for the means of financing investments undergo cyclical changes. The change in the liquidity preference function, which is central to Keynesian theory, has been studied by Kragh (1955, p. 168). During an expansion period, characterized by growing investment, there will be a large demand for bonds resulting in rising bond prices. When the interest is decreasing, those who have bought bonds can be expected to sell them when the bond prices have risen far enough to match their valuations.

Figure 5.9 shows the net increase in cash, equal to the surplus demand for bonds, which households have had to accept in lieu of bonds. The bond stock is redistributed and a new liquidity preference curve indicates the interest rates, where the new owners plan to sell their stock of bonds. The new curve is more elastic than the old one. A continued increase in the money stock because of a large demand for bonds leads to a stepwise lowering of the market interest rate and change in the liquidity preference curve, which will become gradually more elastic. This is a central hypothesis in Keynesian theory. As this change in the liquidity preference curve is dependent on the growth in investment, it can be said to be a function of the growth of production as this is dependent on the change in investment because of the multiplier.

The sources of finance will also depend on income growth rate. In Kragh's model it is assumed that current investment is completely financed by bank loans. It is also assumed that during the following period these bank loans are consolidated by bond flotations. If part of the current investment is financed by firms' own savings the supply of bonds for refinancing these bank loans will be reduced to the same degree. As the firms' profit share, and thus their savings, varies during the cycle as it can be assumed to be a function of the firms' capacity utilization, the supply of bonds will be dependent on capacity utilization and income growth rate.

Kragh (1955, p. 167) has shown that the condition for stability in his model is that the slope of the liquidity curve is higher than the slope of the investment curve. The formal presentation of his model includes an investment function:

$$a_1 - b_1 I_t = R_{t-1}$$

$$\Delta N = \Delta I$$

where ΔN is the increase in the public's cash holding, R_{t-1} is the rate of interest in the preceding period, $R_t = a_2 - b_2 N_t$ and N_t = the public's cash holding.

The solution of the model with regard to the rate of interest is (K_1 and K_2 are constants)

$$R_1 = K_1 \left(\frac{b_2}{b_1} \right)^L + K_2$$

If $I_t > I_{t-1}$, so that there will be surplus demand for bonds and hence a lower interest rate, an equilibrium will be reached if $b_1 > b_2$, i.e. $\partial I / \partial R > \partial N / \partial R$, which implies that the slope of the liquidity curve is greater than the slope of the investment curve.

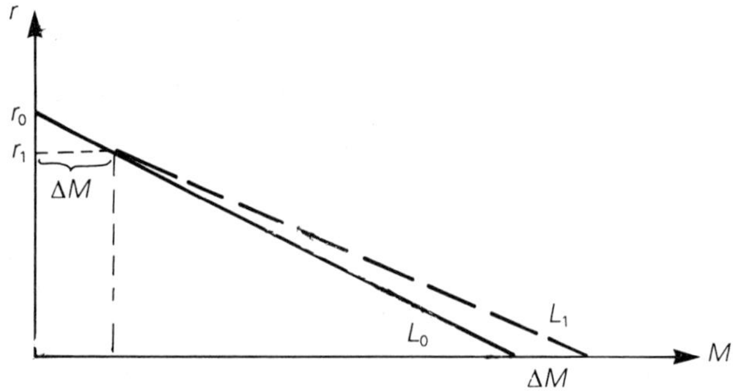

Fig. 5.9 Liquidity preference curve with increasing money stock.

This result will not be changed by the movements of the demand and supply functions which, as already noted, are dependent on the income growth rate. In fact, changes in the curves as a result of income growth will tend to make the liquidity curve more elastic. This will also be the result of a decreasing profit share during income expansion, which will cause an increasing bond supply in order to consolidate bank loans. However, this increase in liquidity will be limited by the value at a profit share of zero, i.e. the value assumed in Kragh's original example. The introduction of business savings in the supply function for bonds would be formulated as follows:

$$a_1 - b_1 I_t - c_1 I = r_{t-1}$$

where c_1 is the share of investment financed by business saving. Therefore the introduction of business savings will substantially increase the elasticity of bond supply with regard to the interest rate. The stabilization of the growth rate will cause a decrease in the profit share and thus in business savings. However, even if it decreased to zero the supply elasticity $\partial I/\partial r$ should be greater than the demand elasticity, $\partial M/\partial r$, which leads to a stable equilibrium. Thus the assumption of the stability of equilibrium in Kragh's model will not be contravened.

There will also be a change in the liquidity preference function when production and investment decline during recession. According to the model the decrease in investment will produce a supply surplus in the bond market and result in a rise in the market interest rate. The fall in bond prices will stimulate speculators to buy bonds when prices have fallen to a level equal to their buying plans. This is indicated by the negative part of money demand on the liquidity preference curve (Fig. 5.10). The redistribution of the bond stock moves the liquidity preference curve, which shows at what market interest speculators are willing to sell their bonds, to the left. It also changes the slope of the curve. It is now less elastic, which means that during a continuous supply surplus of bonds the coefficient of the liquidity

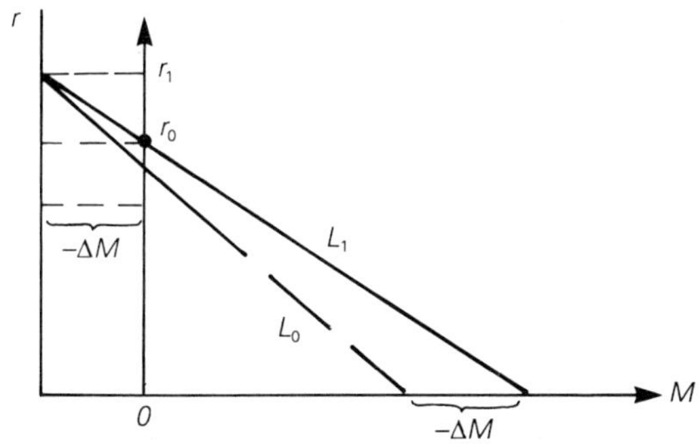

Fig. 5.10 Change in the liquidity preference during a recession.

preference function b_2 will rise. As we have seen that a reduction in the profit share may result in an increase in the need for external finance, there may be an increase in the desired bond supply. The effect would be a decrease in the investment coefficient b_1. Therefore the difference $b_1 - b_2$ may be reduced, which would result in a slower convergence process. b_1 may even be smaller than b_2, which means an unstable situation. Changes in the income level would tend to depart from equilibrium.

(b) *The introduction of equities in the Keynesian liquidity preference theory*

Among the sources of finance available to enterprises, equity finance is particularly important as it is necessary for enterprises to keep a stable relation between debt and equity capital in the medium and long term. This also means that financial investment in households and credit institutes should include equities.

In the transactions table (Fig. 5.11) there will be at least three types of financial assets in the portfolios of households and the sector for credit institutes outside banks, the so-called 'other credit institutes' (OCIs). The banks are assumed to keep their assets in loans to the public and in bonds. As assets and debts inside the sector are netted out, equities and bonds are registered in the non-financial enterprises only if they are issued in the foreign sector or in credit institutes. However, we shall still assume a closed economy, and so the transactions table will only contain four sectors, non-financial enterprises (NON), banks (BAN), other credit institutes (OCI) and households (HOU).

The disequilibrium analysis of the financial markets can be developed using the demand functions for assets introduced by Benjamin Friedman in his studies of the income effects of budget policy. Like Kragh, Friedman assumes a closed economy where private wealth includes three assets: money, bonds and real capital. Another resemblance is that goods prices are

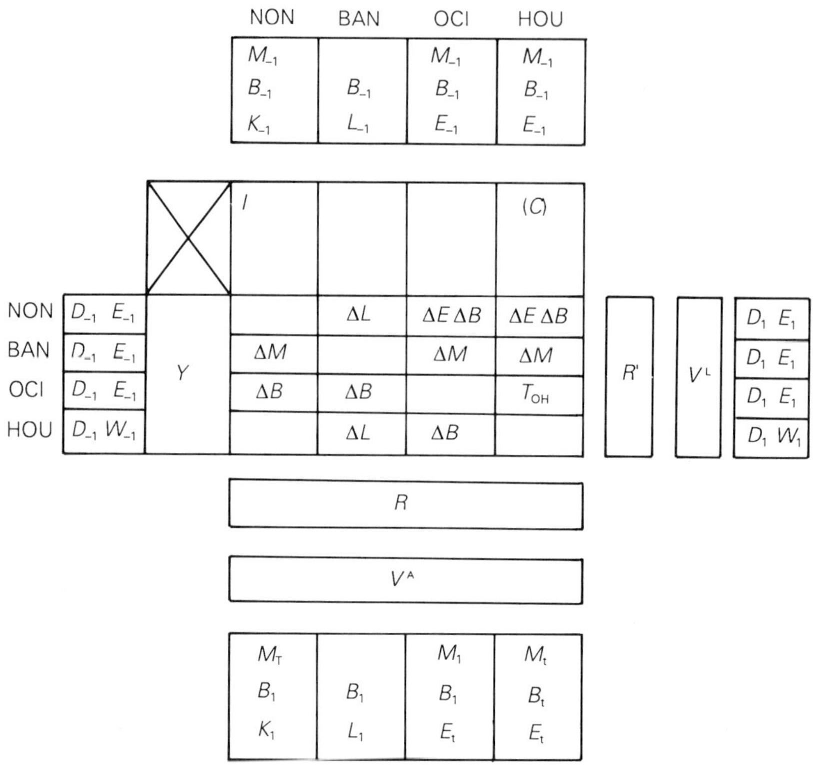

Fig. 5.11 Transactions table for a financial model including equities: B, bonds; C, consumption; D, debt; E, equities; I, investment; K, real capital; L, bank loans; M, money; R, resources; T, transfer; V^{A}, revaluation of assets; V^{L}, revalution of liabilities; W, wealth.

assumed to be constant, so that there is no need to distinguish between real and nominal magnitudes.

The demand functions for assets in Friedman's model are written

$$\begin{bmatrix} M^{D} \\ B^{D} \\ K^{D} \end{bmatrix} = \begin{bmatrix} m_0 \\ b_0 \\ k_0 \end{bmatrix} + \begin{bmatrix} m_1 & m_2 & m_3 \\ b_1 & b_2 & b_3 \\ k_1 & k_2 & k_3 \end{bmatrix} \begin{bmatrix} r_M \\ r_B \\ r_K \end{bmatrix} + \begin{bmatrix} m_4 \\ b_4 \\ k_4 \end{bmatrix} Y + \begin{bmatrix} m_5 \\ b_5 \\ k_5 \end{bmatrix} W$$

where M^{D}, B^{D} and K^{D} are the amounts demanded of money, bonds and real capital respectively, m_i, b_l, k_i are fixed coefficients, r_M is the known yield of money, and r_B and r_K are the expected yields on bonds and real capital respectively. It follows from the balance-sheet constraints that

$$m_j + b_j + k_j = 0 \qquad j = 0, \ldots, 4$$

$$m_5 + b_5 + k_5 = 1$$

The following symmetry constraints are imposed on the asset demand functions:

$$b_1 = m_2 \qquad k_1 = m_3 \qquad k_2 = b_3$$

In Friedman's model the three assets are assumed to be gross substitutes. Karacaoglou (1984) has criticized Friedman's assumptions about gross substitution on the grounds of incompatibility with Keynesian theory. His criticism, which is justified if real capital is included among the assets, loses its relevance if real capital is replaced by equities.

The assumption of gross substitution implies the following restrictions on the coefficient of the asset demand function in Friedman's model:

$$m_2 = b_1 < 0$$
$$m_3 = k_1 < 0$$
$$b_3 = k_2 < 0$$

Karacaoglou has instead the following two sets of restrictions:

$$m_2 = b_1 < 0$$
$$m_3 = k_1 = b_3 = k_2 = 0$$

and

$$m_4 = -b_4 \qquad \text{i.e. } k_4 = 0$$

He comments on this with a quotation from Davidson (1980, p. 315): 'substitutability between money and liquid assets (but not *illiquid durables*) is the essence of Keynes' liquidity preference theory'.

For our purposes it is necessary to assume that the portfolio owners do not have direct access to the real capital in their portfolios, but instead keep among their assets the equities of firms who use the real capital in the production process. Therefore real capital is substituted for equities in the demand functions. In our model the demand functions for assets can be formulated in the same way as in Friedman's model, except for one important difference: the influence of asset prices on demand is in turn dependent on the development of production. We therefore choose a multiplicative system rather than the additive system adopted by Friedman. We assume that the relations have a constant elasticity form:

$$M^D = m_0 r_M{}^{m_1} r_B{}^{m_2} r_E{}^{m_3} Y^{m_4} W^{m_5}$$
$$B^D = b_0 r_M{}^{b_1} r_B{}^{b_2} r_E{}^{b_3} Y^{b_4} W^{b_5}$$
$$E^D = e_0 r_M{}^{e_1} r_B{}^{e_2} r_E{}^{e_3} Y^{e_4} W^{e_5}$$

In logarithmic form the model is written

$$\ln M^D = \ln m_0 + m_1 \ln r_M + m_2 \ln r_B + m_3 \ln r_E + m_4 \ln Y + m_5 \ln W$$

$$\ln B^D = \ln b_0 + b_1 \ln r_M + b_2 \ln r_B + b_3 \ln r_E + b_4 \ln Y + b_5 \ln W$$

$$\ln E^D = \ln e_0 + e_1 \ln r_M + e_2 \ln r_B + e_3 \ln r_E + e_4 \ln Y + e_5 \ln W$$

As we have no information about the equity stock the function for $\ln E^D$ is written

$$\ln r_E = \ln E^D - \ln e_2 - \frac{e_1}{e_3} \ln r_M - \frac{e_2}{e_3} \ln r_B - \frac{e_4}{e_3} \ln Y - \frac{e_5}{e_3} \ln W$$

where $\ln E^D - \ln e_2$ is a constant. As the equity issues are relatively small relative to the equity stock, this form of the equation will probably not introduce any major errors.

(c) *Conclusion*

Summarizing the hypotheses of Friedman and Karacaoglou we can say that their three central relations, which Friedman assumes to be negative, refer to money in relation to bonds, money in relation to equities and bonds in relation to equities (Fig. 5.12). The first relation, which refers to the Keynesian liquidity preference, is also negative according to Karacaoglou's hypothesis. There is no connection between the other relations. However, it should be remembered that Karacaoglou relates money and bonds to real capital and not to equities.

According to our theory the presence of these three relations depends on the production growth rate. When the growth rate is high there is a complementary relation between the three types of assets. The demand curves tend to be horizontal as shown by the broken lines in Fig. 5.12. When the growth rate slows down a competitive relation arises between the three types of assets in the asset-holders' portfolios, and the asset prices become important for asset distribution. Under these conditions the demand curves have a negative slope.

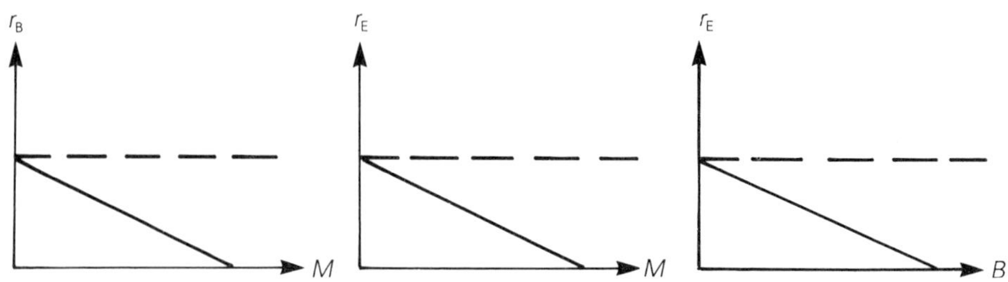

Fig. 5.12 Effect of changing production growth rate on demands for financial assets.

(a) *Relations between production and financial development: cyclical phases*

(i) Phase 1. In previous chapters we have analysed how production development influences costs and prices and the factor income distribution. It has been shown that technical development changes the marginal cost in industry so that an increasing capacity utilization during the phase of cyclical expansion leads to decreasing costs. The result will be a change in factor income distribution in the direction of an increasing profit share. This cyclical development characterizes what has been called phase 1. The situation is illustrated in Fig. 5.13.

Phase 1 is the first phase of cyclical expansion. There is a high production growth from a level of fairly low capacity utilization. Therefore there is only a low and decreasing inflation rate. There is a balance-of-payments surplus since exports are expanding and imports have not yet started to grow very much. There is a strong growth in the profit share as a result of the growth in productivity.

This phase is generally characterized by an increase in liquidity in industrial countries. The balance-of-payments surplus is a result of an international improvement in the production growth rate. A general easing in the financial situation with better internal and external equilibrium will lead to lower market rates on the long bond market. Also the short bond

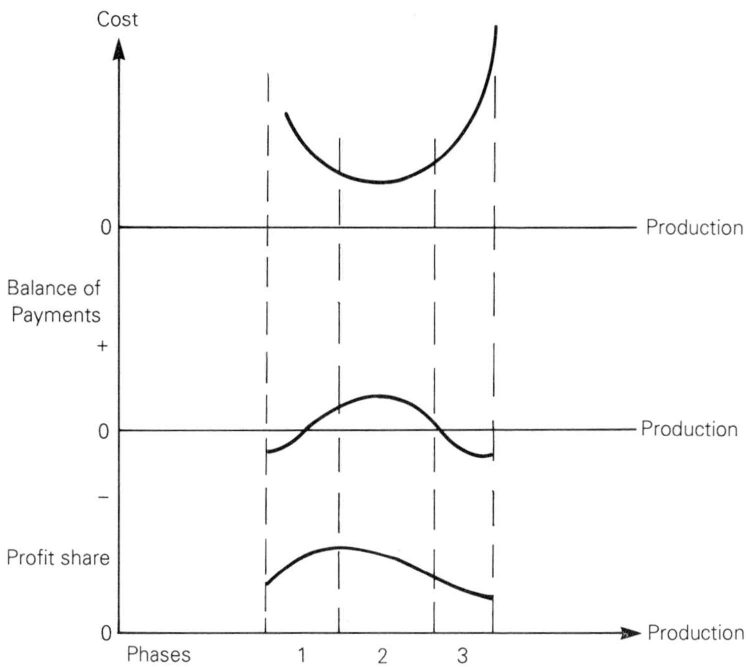

Fig. 5.13 Production development, balance of payments and the profit share.

rate may decrease as the exchange reserves are increasing as a result of the balance-of-payments surplus and the central banks can adopt a more liberal credit policy as the inflation rate is decreasing.

The development of the equity market will be influenced by the high profit growth which leads to an increasing profit share. The result will be a rapid increase in equity prices stimulated by increasing expectations of income growth. At the same time there will be a low growth in equity issues as investment growth is low as long as there is still a fairly low, although increasing, capacity utilization.

The financial development during phase 1 is influenced by a high liquidity growth and a high profit growth which leads to a high complementary demand for all types of financial investments in order to diversify portfolios. The growth rate of prices for equities and bonds will also be stimulated by the low rate of supply in these markets as investment in real capital has a low growth rate and the increase in income growth has increased tax incomes in the government sector and leads to improvements in the budget balance.

The disequilibrium thus created can be observed in the transactions table by the difference in the value of the assets of equities and bonds in the 'households' and 'other credit institutes' sectors, and the nominal values of these securities among the liabilities in the non-financial enterprises sector where these securities have been issued. In a steady-state growth there would not have been any difference between the valuations of these securities from the asset side and from the liabilities side. In a period of expansion the valuation from the asset side will be higher than the valuation from the liability side. These changes in asset prices are registered in the matrices of capital valuation: V^{A} for the asset side and V^{L} for the liability side. The development will influence the demand for final products in both the non-financial enterprises sector and the households sector.[1]

(ii) Phase 2. In phase 2 the economy has reached its normal long-run capacity utilization. As the production growth rate is still high there is now need for a higher growth in production capacity. Therefore the growth in investment increases. In this phase the profit share begins to decline as the wage share increases as a result of both increasing employment and higher wage rates. Although the rate of inflation is rising, the growth rates of prices and productivity are not equal to the growth in labour costs. This can be seen from the development of the supply curve in Fig. 5.13 during phase 2. The figure also shows that both the balance of payments and the profit share peak in this phase. The development of the balance of payments, in particular, is a result of the increasing growth rate of imports which is a

[1] Taylor and O'Connell (1988) have assumed market values for the equity capital among the firms' liabilities, although they study adjustment dynamics.

consequence of the high capacity utilization, which results in demand for scarce goods and services from foreign markets.

The increasing growth in investment in real capital will keep the production growth rate stable. The financing of capital building must now be met with a higher share of external finance as the business sector's own savings are not growing sufficiently rapidly. However, this can be done without increasing interest rates as the higher wage share produces an increase not only in the lending capacity in the household sector but also in the capacity of the various capital market institutes and funds whose resources depend on income growth in the household sector. The strong expansion of the role of these types of institutes in many industrial countries has increased the tendency to complementary demand for different types of financial assets in times of high production growth. This development is stronger in phase 2 than in phase 1 partly because of the higher growth rate of external finance and partly because of the change in factor income distribution which leads to a higher growth rate for financial resources in capital market institutes and investment funds.

It is therefore quite probable that the growth rate of equity prices will be stable as the decrease in the profit share will be balanced by an increasing growth in demand from capital market institutes and investment funds. The expectations of investors will be very sensitive to changes in the GNP growth rate. A stable growth rate for the GNP will help to keep up the growth rate for equity prices. A risk may be the development of the balance of payments. Its surplus is assumed to have peaked during this phase. However, as long as there is a surplus it will provide an inflow of liquidity which will stimulate lending by the banking system. Another risk may be an increasing inflation rate. If this has reached such a level that it threatens the stability of the growth of wage rates and the competitive position of the export industry, the central bank may introduce a restrictive monetary policy. In such a case short bond rates may rise even if long bond rates remain stable as they are more dependent on the international business cycle. The risk is that an increase in the short bond rate may influence equity market prices. However, we believe that stable growth rate during this phase will have a stronger influence on the prices of financial assets than the tendencies to increasing prices as long as there is an inflow of liquidity from foreign markets.

In his analysis of financial development during a long period of stable growth Minsky (1964, p. 213) has emphasized how the increased need of external finances can be realized as the expectation of future growth of profit is spread in the economy. He notes how these expectations influence the valuations of the assets.

(iii) Phase 3. Phase 3 starts with a fall in the GNP growth rate. This may be due to a decrease in international or home market demand; it may also

be the result of bottlenecks in production capacity, raw materials and energy. Therefore even though capacity utilization may stagnate, there is a strong cost inflation as a result of a decrease in productivity at the same time as wage rates continue to increase because of the institutional time lag between production growth and growth in wage settlements. Thus there is a continuous change in the distribution of factor income leading to a decrease in profit share. The balance of payments becomes negative because household income and consumption, and hence imports, continue to grow while exports decrease because of a lower growth in international demand and a deterioration in the competitive position of the export industry.

There is a strong decrease in liquidity in this phase because of the deficit in the balance of payments. Therefore the credit capacity of the banking sector will decrease and the monetary policy will probably strengthen this restrictive tendency in order to stimulate capital imports. The result will be an increasing short bond rate. This development, together with a decrease in the GNP growth rate which will lead to decreasing income expectations, will result in a fall in the equity prices. If the recession is international there is a risk that there will be an increasing disequilibrium in several countries and the interest rate for long bonds will also tend to rise. Such a development will make the financing of the building industry and the consolidation of bank loans by bond flotation more expensive and more difficult to be taken on by capital market institutes. There will be competition between those in need of external finance as financial investment tends to go down as investors require more liquidity. The tendency to choose between different types of assets means that the complementary demand for various types of financial assets which exists when the growth rate is high is replaced by a tendency to substitute between the different types of assets in the new situation of decreasing growth.

The new disequilibrium situation can be analysed by comparing the flows on the financial matrix. When the non-financial enterprises try to float new issues of bonds or equities in order to consolidate the bank loans which have financed current investment expenditures, they are discouraged by the capital market institutes who 'will attempt to protect their "goodwill" with their normal customers (who have brought previous new issues) by reducing the pressure on security prices' (Davidson, 1978, p. 326). Therefore enterprises will find it difficult to finance their investments. They may have to 'increase their indebtedness to the banking system in order to finance an "undesired" increase in dealer inventories' (Davidson, 1978, p. 326).

The excess flow supply in securities can be indicated in the matrix for financial transactions by a large value for the demand for liquidity in the households and other credit institutes sectors (Fig. 5.14). Therefore the flows of financial investments in these sectors increase for banks (F_{23} and F_{24}) and decrease for non-financial enterprises (F_{13} and F_{14}). The disequilibrium is indicated by the fact that the investment planned by households

	NON_1	BAN_1	OCI_1	HOU_1
NON	–	F_{12}	F_{13}	F_{14}
BAN		–	F_{23}	F_{24}
OCI			–	(F_{34})
HOU				–

Fig. 5.14 Financial matrix.

and other credit institutes is less than the supply of bonds and equities planned by non-financial enterprises. The matrix also shows how the balance can be restored *ex post* by an increase in the credits for non-financial enterprises in the banking sector (F_{12}). Such a development would mean a change in monetary policy in order to facilitate an expansion of bank credit. The main obstacle for such a change in policy would be the balance-of-payments deficit. This cannot be studied in the financial matrix as the foreign sector is not included.

The disequilibrium situation in the financial markets in phase 3 can also be analysed from the point of view of asset valuation. The change in the preferences for financial investment is the result of the decrease in the production growth rate. As long as the growth rate is stable and on the expected level, a decrease in the profit share will lead to a higher uncertainty. However, the possibility of portfolio diversification and of investment in property with a high resistance to inflation, such as real estate, gold, precious stones and works of art, may provoke increased speculation. It is not so much expectations about future profit development as expectations about the opinion of other operators' expectations about tomorrow's equity prices which is the driving force in such a boom. A looming risk may be the fear of exchange depreciation and the increase in the discount rate that the central bank may have to introduce as a countermeasure.

The decline in the production rate will bring falling security prices to the fore. The appreciation which had resulted earlier from an expectation of growing profits disappears. The asset prices in owners' balances will decrease relative to the nominal values of debts and equities in the passive balance of the enterprises who have issued them. Such a development will lead to increased financial difficulties if intermediaries are forced to sell their portfolios when asset prices are falling in order to increase their liquidity. We will obtain what Minsky (1964, p. 262) has called a 'financial crisis'.

(b) *Structural changes and stability*

From the Second World War up to the middle of the 1970s recessions in the industrial countries—the so-called phase 3—had generally been rather short. However, during the last 15 years, there have been prolonged periods of stagnation, which makes it tempting to define a fourth phase characterized by prolongations of the recession occurring during phase 3. However, this development seems to be more a result of a structural change in the economy than a phase in the business cycle.

In Chapter 2 we analysed the results of some structural changes in cost and price development in the industrial countries. It was noticed quite early in the post-war period that the increasing share of the service sector in total production resulted in a decrease in average productivity growth. This tendency was particularly obvious in the European countries which had been industrialized at an early stage (Kaldor, 1966). In these countries the primary production sector, particularly agriculture, had already been reduced to a small share of total employment, so that in times of low population growth the expanding service sector attracted its inflow of labour force mostly from the industrial sector. The result was a decrease in the average rate of growth of productivity in the medium term, as the share of the service sector with its low productivity growth increased relative to the industrial production sector with a much higher productivity growth rate. This development had serious consequences for the ability to control rising costs in a situation of a high growth rate of demand and a high employment level. The cost and supply curves on the macroeconomic level tended to be steeper in the medium term. Figure 5.15 shows how the supply curve changes in the medium term and becomes very steep long before the production level which had earlier been regarded as the full employment level is reached.

This structural change leads to a higher inflation rate when the goals for full employment are established on a low unemployment level. It will also aggravate disequilibrium in the balance of payments. In most countries the decreasing share of the industrial production sector will increase the need for more imports, so that the import ratio will rise. In some countries exports are concentrated in the industrial sector. In this case the structural change may reduce the balance-of-payments surplus by both increasing imports and decreasing exports. Fig. 5.15 shows how the structural change has influenced the development of the balance of payments over the cyclical phases. In an expanding economy the surplus in the first two phases may decrease and a deficit may already be developing in phase 2. This development will be strengthened by the higher growth rate in costs and inflation during phase 2. There is therefore a risk that inflation may be increased further by devaluation in phases 2 and 3.

The deep recession in many industrial countries during the second half of the 1970s led to the closure of enterprises that had become unprofitable because of the increase in costs. These liquidations resulted in a rise in the average productivity for that part of industry which survived the crisis. This

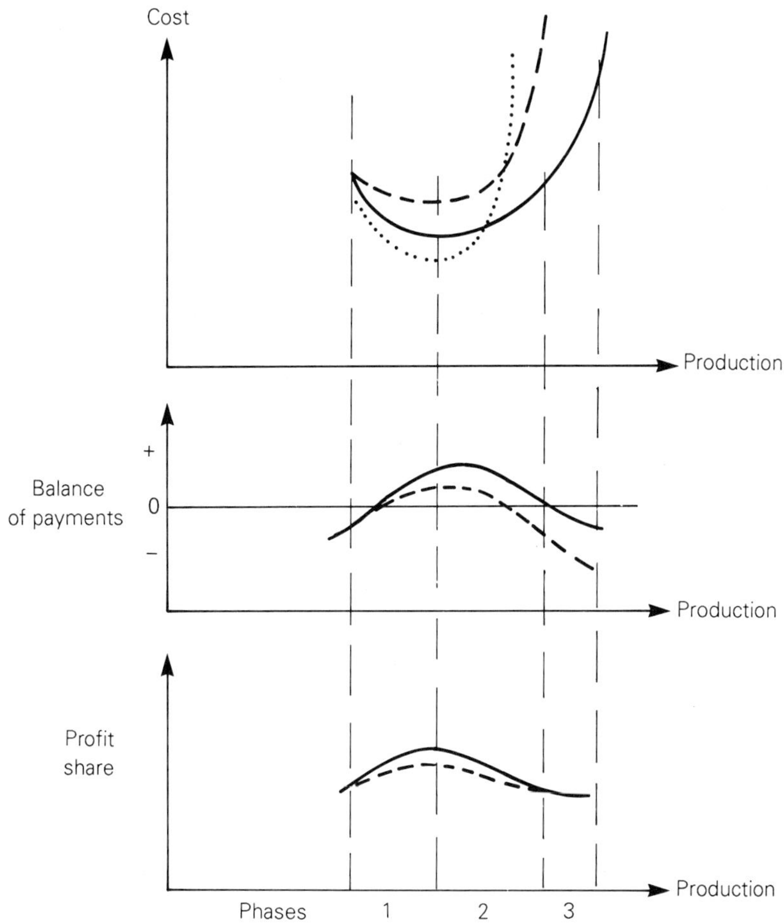

Fig. 5.15 Structural changes in productivity.

development strengthened the tendency to increasing productivity when the capacity utilization is increased (Industriell förnyelse, 1987, p. 13). Therefore the development during the second part of the 1970s led to important consequences for the change in costs over the business cycle: a decrease in costs when capacity utilization is increased from a low level of ultilization, and an increase in costs when the economy is approaching a further growth in capacity utilization. These changes will lead to a new type of cost curve with a more pronouced U shape, as illustrated in Fig. 5.15.

The marked increase in productivity when capacity utilization is increased from a low employment level—the Verdoorn effect—has been frequently observed since the Second World War (Kaldor, 1966, p. 10). The rapid increase in costs even before full employment is reached is the result of the

Phillips effect, which has been much studied since the 1950s. Therefore the developments in the 1970s resulted in a reinforcement of both these effects. As there has been a strengthening of oligopolistic tendencies in both the industrial and service production sectors, it is quite possible that a recession is not followed by any substantial price reductions in such a case. In the recessions during the second part of the 1970s there was a substantial reduction in productivity. Thus the cost curve changed compared with the earlier period.

Figure 5.16 shows how the more favourable cost situation is found below the full capacity level Y^* when demand goes down during a recession. It would therefore be possible to find a stable equilibrium on a production level that implied unemployment and a low capacity utilization, which is in agreement with traditional Keynesian theory.

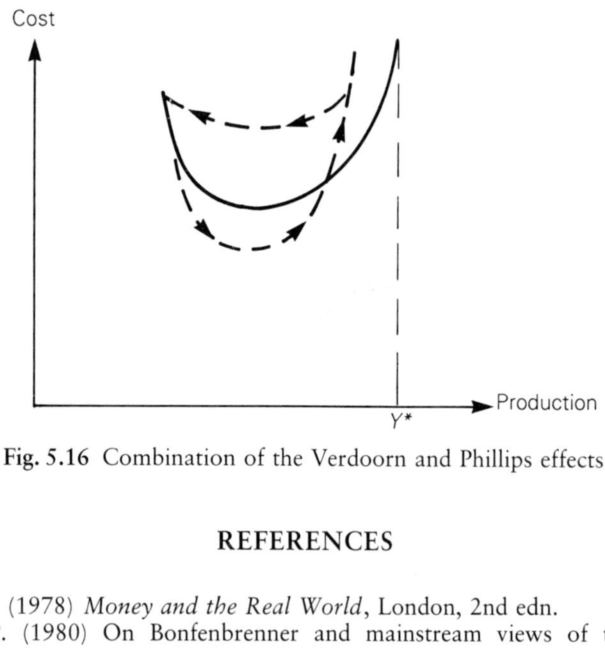

Fig. 5.16 Combination of the Verdoorn and Phillips effects.

REFERENCES

Davidson, P. (1978) *Money and the Real World*, London, 2nd edn.

Davidson, P. (1980) On Bonfenbrenner and mainstream views of the 'essential properties' of money: A reply. *Journal of Post Keynesian Economics*, **2** (3).

Friedman, B. (1978)

Gurley, J. G. and Shaw, E. S. (1960) *Money in a Theory of Finance*, The Brookings Institution, Washington, DC.

Industriell förnyelse (1987) *Statens Offentliga Utredningar*, Liber, Stockholm.

Kaldor, N. (1966) *Causes of the Slow Rate of Growth of the United Kingdom. An Inaugural Lecture*, Cambridge University Press, Cambridge.

Karacaoglou, G. (1984) Absence of gross substitution in portfolios and demand for finance: Some macroeconomic implications. *Journal of Post Keynesian Economics*, 6 (4), 580–1.

Keynes, J. M. (1936)

Kragh, B. (1955) *The Meaning and the Use of Liquidity Curves in Keynesian Theory*, Macmillan, London, International Economic Paper No. 5.

Kragh, B. (1967a) *Finansiella Långtidsperspektiv*, Esselte AB, Stockholm.

Minsky, H. (1964) Financial Crisis: financial systems and the performance of the economy, in *Private Capital Markets* (eds I. Friend, H. Minsky and V. Andrews), Prentice-Hall, Englewood Cliffs, NJ.

Taylor, L. and O'Connell, S. A. (1988) A Minsky crisis. *Quarterly Journal of Economics*, C, Suppl., 871–85.

6

Economic policy and economic growth

6.1 FINANCIAL DEVELOPMENT DURING A PERIOD OF HIGH ECONOMIC GROWTH

6.1.1 The high liquidity inherited from the Second World War

The enormous expansion in government expenditure in many industrial countries during the Second World War led to a financial expansion. In the Anglo-Saxon countries and also in some countries which did not take part in the war this financial expansion did not cause an immediate inflationary process. In fact it had been possible to keep demand down by a restrictive policy. Banks and other credit institutes had financed most of the government debt. In the United Kingdom, for instance, the government financed its budget deficit mostly by issuing treasury bills and short- and medium-term bonds. The banking system had to redeposit its surplus of investable funds with the Treasury, thereby limiting the amount of credit that could be extended to the private sector (Kaldor, 1982).

The enormous increase in the public debt in these countries had been created at the same time as inflation had been increasing only moderately, and the interest rate had been maintained at an extremely low level. In fact the interest rates had been low ever since the United Kingdom had abandoned the Gold Standard in 1931. In both the United Kingdom and Sweden there was even reduction in the interest rates in the later part of the 1940s.

In Sweden, the expansion of demand immediately after the War led to a strong increase in imports, particularly from 1946 to 1947. Liquidity in the banking system decreased, but credit expansion was able to continue as the cheap money policy forced the central bank—the Riksbank—to buy the bonds offered by the banking system. The banking system sold most of its stock of bonds built up during the war years. It was reduced by about a thousand million crowns.

The cheap money policy in Sweden had both theoretical and practical political foundations. The Keynesian theory for the need of a low interest rate was accepted and a cyclical recession was expected in the near future. In order to offset the decrease in demand by house building financed by local government, it was essential to maintain a low interest rate.

After the outbreak of the Korean War in 1950 this policy became more difficult to implement. Rearmament, particularly in the United States, led to rising raw material prices, and in the forest industries, for instance, there was a strong increase in liquidity and profits. It was necessary to use new regulations to keep down bank credits. The banks were obliged to keep up to 50% of their deposits in cash and to hold other liquid assets as foreign exchange and bonds. These measures seemed to have little effect, and when there was a further increase in demand in the middle of the 1950s the interest rates were allowed to rise.

In the following years of high growth rate in most industrial countries during the 1950s and 1960s changes in interest rate were used to influence demand. In particular, in the years of cyclical expansion increases in the interest rate were used to offset the decline in exchange reserves resulting from the high import demand. In many countries, however, bank regulation was used at the same time to direct the financial flows on the credit market.

It is remarkable how such regulations remained in Sweden, for example, when restrictions in foreign trade disappeared during the 1950s and the 1960s. The reason was that even if the internal and external balance of this country underwent cyclical fluctuations, in general the balance was characterized by demand surplus during most of the period of relatively high growth rate in the 1950s and 1960s. This was also the situation in most West European industrial countries in this period.

The surplus demand in an open economy with private and government sectors can be studied in a transactions table. This will first be constructed for an *ex ante* analysis of an economy with four sectors: non-financial enterprises, banks, government and households. The transactions table for the *ex post* analysis after economic and political measures have been applied is constructed subsequently. The combined table is shown in Fig. 6.1.

The transactions table *ex ante* registers the surplus demand for finished goods for both capital building and consumption. However, this demand surplus is not allowed to lead to rising prices. Prices and wages are controlled. Investment is kept down by means of building regulations and taxes are increased in order to reduce household disposable income and to keep down consumption demand to what is believed to be the planned production level of consumption goods and services. Firms' investments are also kept down by restrictions on financing. There is both control of bond issues and restrictions on bank credit. Therefore house-building and government sector bond issues have preference in bank portfolios according to agreements between the banks and the authorities.

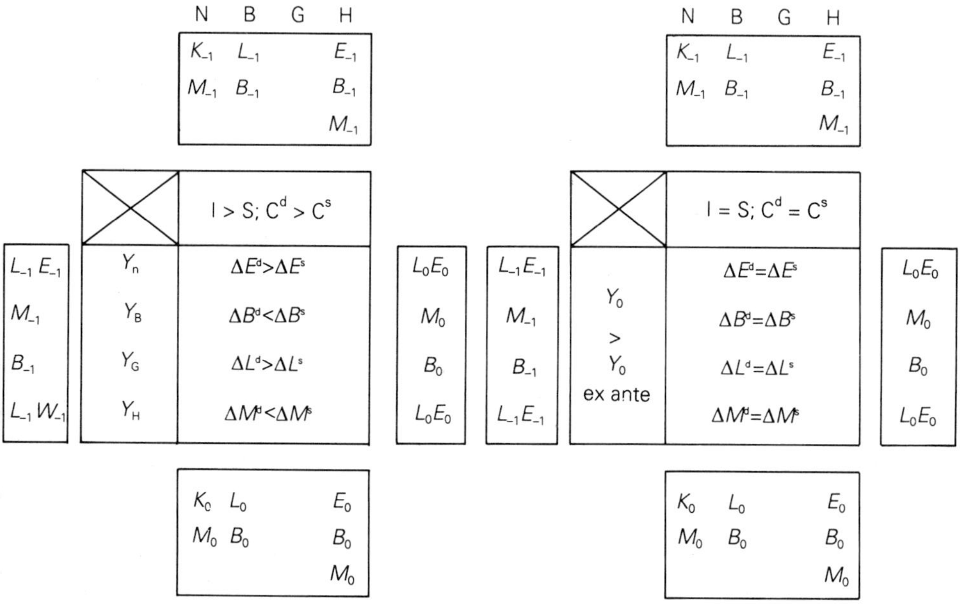

Fig. 6.1 Transactions table with surplus demand and restrictive economic policy.

		Economic policy
$C^D > C^S$ $I > S$	$C = C^S$ $I = S$	C^D reduced by increasing taxes I reduced by building restrictions
$\Delta B^S > \Delta B^D$	$\Delta B = \Delta B^D$	ΔB^S restricted by controls of issues
$\Delta E^D > \Delta E^S$	$\Delta E = \Delta E^S$	ΔE^D restricted by dividend control
$\Delta L^D > \Delta L^S$	$\Delta L = \Delta L^S$	There are also wage and price controls
$\Delta M^S > \Delta M^D$	$\Delta M = \Delta M^S$	

As prices are not allowed to move relative to the disequilibrium in markets for goods, services and assets, the matrices for valuation changes do not register the price changes that should have been the result of the disequilibrium and the changes in total value and distribution which should have been the result of such price changes are not included in the outgoing assets.

However, the long period of expanding demand in the industrial countries during the 1950s and 1960s was interrupted by short recessions. The first came in 1952–3 after the expansion produced by the outbreak of the Korean War. The next recession took place in 1957 and was much more severe, particularly in the United States where the decrease in GNP between 1957 and 1958 was 9%. There were also recessions during the 1960s, but they were less severe.

In most Western European industrial countries the slackening of demand was counteracted by a government policy to stimulate demand. In Sweden, for instance, increasing government expenditure led to budget deficits during the recessions in the years 1953–4 and 1958–9. As the markets for long credits could not absorb the increase in government debt, much of the state borrowing had to be taken over by the banks. Even if these credits were 'locked into' bank assets by agreements between the banks and the central bank, the resulting expansion in income led to an increase in bank liquidity. In the following cyclical rise in demand the banks were able to increase credits for investment and consumption, which resulted in inflationary development.

Swedish economists have interpreted this development according to the concepts of quantitative theory (Lundberg, 1963, pp. 42, 54). The surplus liquidity in the banking system leads to a credit increase which results in a money stock which is 'too large' in the private sector and threatens to produce an expenditure expansion which is too strong in the latter part of the business cycle. Whether the result is an increase in the production volume or in the growth of inflation depends on the degree of capacity utilization. The time lag between the increase in the money stock and the expansion of expenditure and price growth rate is analysed as changes in the velocity of money.

The increased liquidity in the phase of slackening demand together with an increasing wage share in income distribution stimulated investment in house-building. As long as there was no lack of production capacity the income velocity could increase without increasing market interest rates. However, when the increase in expenditure results in income and price increases the rising money velocity eliminates the surplus liquidity and interest rates will increase (Table 6.1).[1] Two alternative definitions of the money stock are used in Table 6.1: (a) banknotes plus demand deposits of the money stock on past accounts; (b) banknotes plus deposits on all accounts in all types of banks. The velocity of money has been calculated as (i) the ratio of the national income to money as defined in (a) and (ii) the ratio of the national income to money as defined in (b). The interest rate is for long government bonds in each country.

[1]Lundberg (1963, p. 43) has compared the growth rate of the national income, the money stock, the velocity of money and the interest rate for bonds in the United States, the United Kingdom and Sweden for the period 1954–7.

Table 6.1

	National income	*Money stock*		*Velocity of money*		*Interest rate*
		(a)	(b)	(a)	(b)	
USA	21.7	5.6	7.9	15.4	12.7	37.7
UK	22.0	2.0	2.8	19.6	18.8	32.8
Sweden	25.2	9.3	17.0	14.6	6.8	33.6

Lundberg interprets the results of the calculations as a confirmation of the hypothesis that the braking effect of the restrictive monetary policy which restrains the increase in liquidity relative to the increase in demand, will be neutralized by the increasing velocity of money. This is a result of considerable elasticity in the provision of means of payment.

Critics of the quantity theory of money have seen the elasticity of the velocity of money as a confirmation of the unreliability of this theory. Kaldor quotes from a footnote in the Radcliffe Report (1959, para. 391) 'that the more efficient the financial structure, the more can the velocity of circulation be stretched without serious inconvenience being caused'. Kaldor's comment on this text is worth reproducing in its entirety:

> I wonder whether the members of the committee were fully aware that in one sentence, or part of a sentence, they repudiated in one fell swoop the quantity theory of money in all its versions, from Cantillon and Hume, through Ricardo, Marshall, and Walras, Irving Fisher and Milton Friedman (to mention only the most prominent) with their army of camp followers right down to Mrs. Thatcher. (Kaldor, 1982, p. 9)

Lundberg's example of the development of income, money stock and the velocity of money in the USA, the UK and Sweden between 1954 and 1957 can be used to study how the mobilization of extra liquidity, which is the meaning of an increasing velocity of money, has resulted in rising interest rates. If we confine ourselves to comparing money and velocity growth according to the wider definition of money, we find that in Sweden it was necessary to have an interest rate growth of 33.6% in order to obtain a growth of 6.8% in the velocity of money. In the United Kingdom it was possible to increase the velocity of money growth by 18.8% with a similar increase in the growth of the interest rate. The higher elasticity in the UK credit system compared with that in Sweden is due to a more developed asset structure with a richer variety of financial opportunities.

During the 1960s there was an important development in the capital market in Sweden. In 1959 a new public pension system was introduced. Central funds were created for the employers' charges. In 1960 pension contributions were 3% of the firms' wage costs, but increased in subsequent years up to 9.5% in 1969. Already in 1965 the funds had reached 10,500 million crowns. The rules for investment of the pension funds made almost all the funds available for the bond market, i.e. for the government and house-building sector, as private bonds were of little importance and were controlled by the central bank. Later a special fund was created inside the pension fund system with the right to make investments in equities.

During the 1960s 'Allmänna Pensionsfonden', which is the official name of the public pension system, was of particular importance for the financing of the house-building programme in local government management. The programme led to the construction of a million new residences and was very important for local development in Sweden, as it was concentrated in the three largest urban areas, Stockholm, Göteborg and Malmö.

6.1.2 Monetary policy in a regulated credit market

Financial development during the 1950s and 1960s indicated that the maintenance of a high production growth rate required a structural change in the financial system. In Sweden the resources in the capital market, in particular, needed reinforcing. In almost all countries the high production growth rate was threatened by disequilibrium in the balance of payments when foreign trade and international financial relations expanded during the 1960s.

Although international relations will be studied more thoroughly in the next chapter, monetary policy cannot be analysed without including foreign transactions. We need a model which, in addition to the sectors for non-financial enterprises, government and households, also includes a foreign sector. The sector for financial enterprises will be divided into a central bank, other banks and other credit institutes. A transactions table for assets, debts and financial transactions (Fig. 6.2) can show the variables which are of importance for the central bank's monetary policy in European countries with a regulated credit market (exemplified by the Swedish experience during this period).

The opening assets are registered above the financial matrix and the closing assets below. The opening liabilities are registered to the left of the financial matrix and the closing liabilities to the right. Only those variables which may be of interest to the monetary policy of a regulated economy are registered as assets or liabilities.

Only one type of asset and one type of debt are registered in the foreign country. The assets can change when non-financial enterprises make loans in foreign countries as an alternative to bank loans in the home country. The foreign sector's debt represents the reserve of foreign assets in the central bank. Maintaining this reserve at a sufficient level can be regarded as the main objective of the central bank. Its instrument is an adjustment of

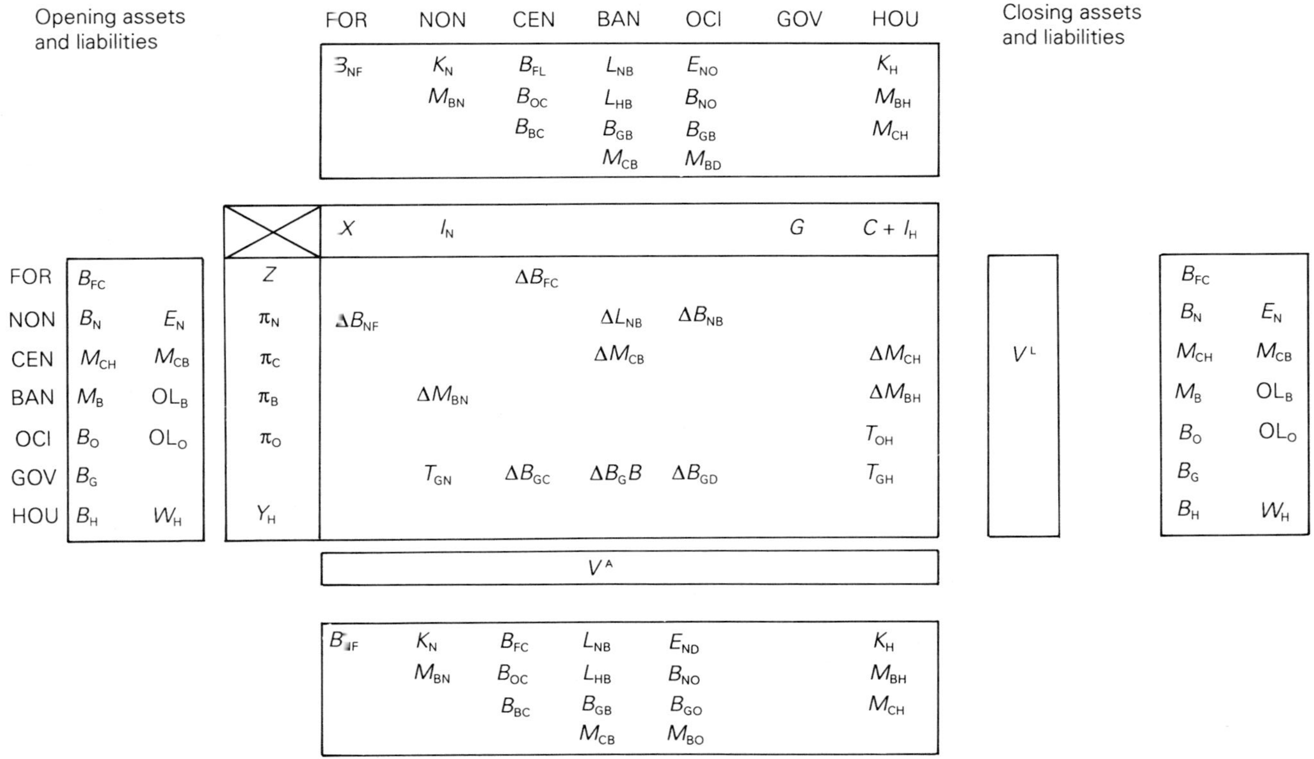

Fig. 6.2 Transactions table for assets, debts and financial transactions.

its discount rate to the short international interest rate. In the 1950s and 1960s the regulation of the bank credits was generally accomplished by fixing the ratio of the bank's cash and liquid assets to its liabilities. In the case of other credit institutes the ratio of the capital market institutes' current investment in priority assets—government bonds, mortgage corporation bonds and mortgage loans—to their total current financial investment was fixed. Another import instrument was the central bank's control of bond issues.

Open market operations did not become an important instrument for the monetary policy because most of the total bond stock—32,000 million crowns out of a total value of 42,000 million crowns—was held in the portfolios of the credit institutes, and their portfolios could be controlled with the regulations described above. This means that during this period the substitution between money and bonds was extremely insignificant in the private sector (Lybeck, 1975, 1978).

The negative relation between the interest rate and the demand for money, commented upon by Lybeck, may be due to borrowing in foreign markets by corporations in need of deposits which results in an increase in foreign currency reserves. Lybeck notes that the central bank cannot sterilize an inflow of foreign exchange, since it has effectively abandoned control over interest rates by setting a rate target (Lybeck, 1975, p. 393).

The household sector's main alternative to money deposits is investment in house building. In the transactions table investment in house building is therefore registered together with consumption in the household sector's column in the matrix for finished goods.

A disequilibrium model for bank credits

Lybeck (1975, p. 301) has analysed a disequilibrium model for bank credits which can illustrate some problems connected with the effects of a regulated credit market in a period of high growth rate. Economic development is characterized by a high profit growth rate which results in a high growth for internal finance in the business sector. Therefore demand for bank loans does not grow as much as the banks' lending capacity.

The discount rate of the central bank is adjusted to the international level of interest rates in order to keep capital import and reserves of foreign currency at a satisfactory level. As the interest rate in the commercial banks is in turn adjusted to the discount rate in the central bank, it is independent of supply and demand for loans in these banks. The supply of loans depends on the growth of deposits and of the rules for the commercial banks' reserves in the central bank, which are used as an instrument to influence their supply of credit. With this institutional restriction the development on the bank loan market can be illustrated using Lybeck's model (Fig. 6.3).

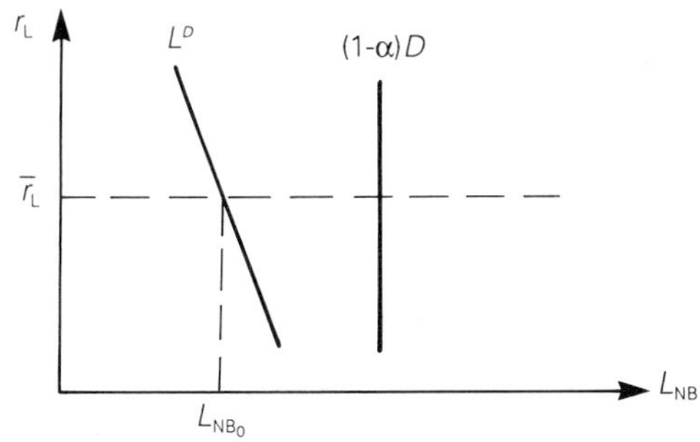

Fig. 6.3 Excess supply of bank loans.

In Fig. 6.3 L_{NB} is the stock of bank loans in the sector for non-financial enterprises, r_L is the bank loan interest rate and $\bar{r}_L$ is the level adjusted to the central bank's discount rate. L_{NB_0} indicates the level of bank loans, which is a result of the demand for bank loans at the given bank rate r_L. In Lybeck's analysis it is assumed that the private sector has a demand function for total external debt, part of which is met by supply from capital market intermediaries. If commercial banks have sufficient liquidity and are not constrained by a loan ceiling, they will satisfy the residual demand (Lybeck, 1975, p. 393).

The supply of bank loans $(1 - \alpha)D$ depends upon the deposits D, and the value of the bank's cash ratio M_{CB}/D (denoted α). As the cash ratio is controlled by the central bank, this bank can also control the commercial banks' supply of loans. The central bank's policy not to reduce the commercial banks' supply capacity by increasing the cash rate can be explained by its wish to make the banks take over government bonds, for which there was not sufficient investment capacity in the capital market institutes. Therefore the supply surplus at the given interest rate would be invested in government bonds.

The model can also be used to illustrate the situation which could develop on the cyclical peak when the commercial banks have expanded bank credit too much to satisfy demand or the central bank has increased the commercial banks' cash ratio in order to reduce bank credits and economic activity (Fig. 6.4). In this situation the amount of bank loans L_{NB_0} will depend on the banks' supply, and there will be a demand surplus of potential credit need which may reduce investment and consumption expenditure. According to Lybeck this situation did not occur in the 1950s and 1960s. The reason for this was that the central bank used other kinds of restrictions on bank credits instead of higher cash ratios to reduce the growth in demand at the peak of cyclical expansion. Lybeck (1975, p. 305) has studied the central bank's use of the loan ceiling. Such a loan ceiling can be introduced in the model as a restriction on credit supply.

The banks are forced by the loan ceiling to reduce bank loans to L_{NB_0}.

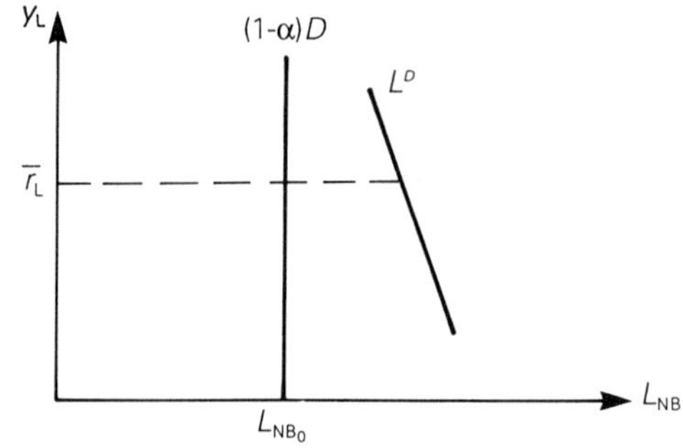

Fig. 6.4 Excess demand for bank loans.

Without this restriction the amount of bank credit is determined by the demand and reaches the level L_{NB_1}. In order to reach the desired credit level L_{NB_0}, the central bank could have increased the cash ratio α instead and so have reduced the credit supply $(1 - \alpha)D$ to the desired level L_{NB_0}. Figure 6.5 shows how the use of the loan ceiling L^s makes it possible for the central bank to obtain another objective in addition to decreasing loans to the business sector. That part of the credit capacity which is not allocated to the business sector has been increased by using the credit ceiling from the value $L_{NB_1} - L_{NB_2}$ to the value $L_{NB_0} - L_{NB_2}$. This credit can instead be allocated to the government or the house-building sector by investments in bonds.[2]

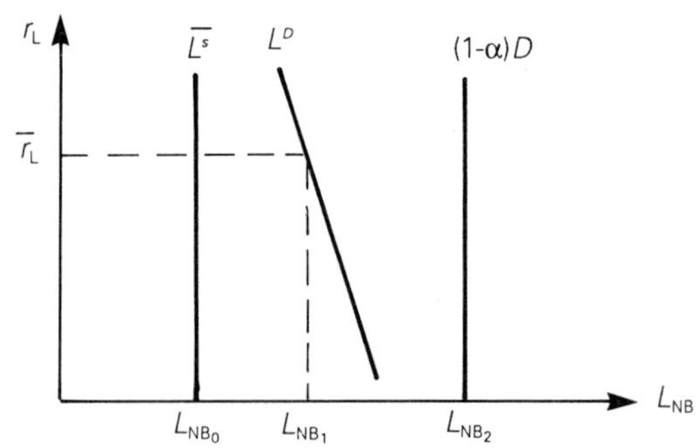

Fig. 6.5 Market for bank loans when there are restrictions imposed by the central bank.

[2]In his analysis of the effects of the credit ceiling Lybeck introduces it in a model with surplus demand. In that case the introduction of the credit ceiling can be seen as reinforcement of a restrictive policy which started with an increase in the cash ratio and led to surplus demand. If, instead, the credit ceiling is introduced in a situation with supply surplus the adoption of this instrument can be seen as an alternative to an increase in the cash ratio. The reason for its use is that the central bank is just as interested in a reallocation of credits from the business sector to the government sector as in a reduction of credits to business.

In his analysis of the use of credit ceilings as an instrument in a restrictive monetary policy Lybeck has stressed its inadequacy as a method of reducing liquidity in the economy (Lybeck, 1975, p. 393). This has been caused by international capital movement. International trade credits free of control have been available in Sweden since the beginning of the 1960s. Whenever the Swedish central bank has increased the discount rate and restricted bank credit, foreign credit has increased liquidity and this has spread through the business sector in the markets outside the organized credit system. Therefore there has been a differentiation of credit conditions such that one part has been controlled by the central bank and the remainder has been governed by the conditions on the grey market. The situation can be illustrated using Lybeck's model for analysing the market for bank loans with credit ceilings (Fig. 6.6).

The model shows how the bank loans at the official bank rate $\bar{r}_L$ have been reduced to L_{NB_0}. However, foreign credits have increased liquidity and created a grey market where an equilibrium exists at the interest rate r_{L1}. The sectors involved in this process can be better illustrated using a transactions table. We shall just study the finance matrix (Fig. 6.7).

The credits ΔB_{NF} of the non-financial enterprises on the international markets are sold to the commercial banks and increase the business sector's deposits by ΔM_{BN}. The foreign exchange is sold by the commercial banks to the central bank, whose exchange reserves rise by ΔB_{FC}. The commercial banks' deposits in the central bank are increased by an equal account ΔM_{CB}. Even if the commercial banks are prevented by the credit ceiling from letting the increased liquidity result in new loans to the business sector, this will just be a credit expansion between enterprises. This will lead to transfers between the different firms' deposits accounts in the commercial banks and so no net change will be registered in the transactions table. However, the intersector credit between the firms, which is assumed to

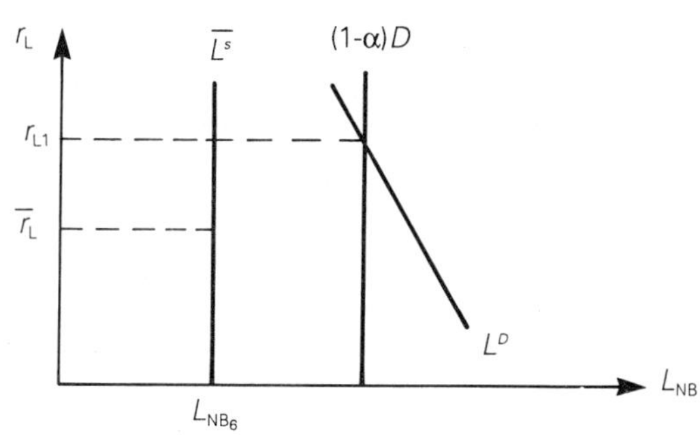

Fig. 6.6 Credit restrictions counteracted by capital imports.

	FOR	NON	CEN	BAN	OCI	HOU
FOR			ΔB_{FC}			
NON	ΔB_{NF}	ΔL_{NN}				
CEN				ΔM_{CB}		
BAN		ΔM_{BN}				
OCI						
HOU						

Fig. 6.7 The finance matrix.

belong to the grey market, can be registered as ΔL_{NN} in the column and row for the non-financial enterprises' sector.

In conclusion we note that at this time the structural change in the credit market had not been adjusted to the financial needs of an economy with a high growth rate, which characterized the development of most industrial countries during the 1950s and the 1960s. For example, the capital market institutes in some countries were not sufficiently developed to handle the financing of the temporary budget deficits resulting from a contracyclical budget policy. In Sweden there was a change in this situation when the state pension system was developed during the 1960s. However, there were still few alternative sources of financing the deficit in either the private or the government sector. Therefore there were few alternative ways of investing the financial savings by households, business and credit institutes. Lybeck (1975, p. 391) asserts that the interest rates do not determine the demand for total deposits, but only the composition of a given stock of money. The reason he gives for this is that there are no financial substitutes in the non-bank private sector. Lybeck notes that it is this sector's ability to shift into and out of real assets that makes the expected rate of inflation the main opportunity cost for keeping money in bank deposits.

It could be said that we have come back to Lundberg's application of the quantity theory in order to explain the effects of financial development on production growth and cyclical instability. As growth in cash deposits was the main substitute both for investment in real capital in the business sector and for house investment in the household sector, bank credit and deposit creation was a fundamental factor underlying production growth and price development. However, it is now easier to see that the application of quantity theory depended strongly on the institutional development of the financial sector.

6.2 STAGNATION AND RISING DISEQUILIBRIUM

6.2.1 Structural change in government expenditure and income

The Second World War changed the role of the government in all industrial countries not only because of the enormous increase in military and other traditional types of government expenditure, but also because of control of private expenditure on consumption and capital building. In many industrial countries, particularly in Western Europe, this control continued after the end of the War. Even in the 1950s when the war damage had been repaired in most areas, there were regulations on trade and payments between many countries. New military campaigns, such as the wars in Korea and Vietnam, and the cold war between the great powers increased military expenditure.

The expansion of the role of government was also a consequence of structural changes in the economy, which led to migration from rural provinces to urban centres. In Europe people moved from the east and the south to the industrial centres in the FRG, Switzerland, eastern France, Belgium and the Netherlands. There was also a redistribution of the working population within countries, such as from northern England to the London area, from the southern to the northern provinces of Italy and from the northern provinces of Sweden to the urban areas around the largest towns.

The result of this structural change was an expansion in government expenditure which was of a long-term or medium-term character and could not be adjusted to cyclical income variations. Changes in government tax incomes could provide better countercyclical income effects than changes in government expenditure. In some countries the construction of the tax systems contributed to automatic adjustments which helped to stabilize income growth. During periods of high growth rate and demand surplus in Sweden the progressive tax system had an automatic income-absorbing effect which stabilized the economy.

After the end of the Vietnam War in the 1970s the growth rate of military expenditure decreased. However, the growth of other types of expenditure kept up the long-run expansion. In Europe the increased growth rate of transfers to households was particularly high. During this decade government expenditure increased, particularly in the FRG and Italy, and the share of government expenditure tended to reach similar levels to shares of GDP (Pedone, 1982, p. 396).

The result of this development was large budget deficits in almost all the industrial countries during the 1970s. When the oil price rises led to falling terms of trade and rising trade balance deficits a restrictive economic policy was introduced in most industrial countries. The result was increasing budget deficits which reached particularly high levels in 1975 after the first oil price crisis. After the second oil price crisis the deflationary efects were prolonged by the very restrictive monetary policy introduced in the United States in 1979. The highest levels of government deficit in the seven major industrial countries were reached in 1982 and 1983, at 3.9% and 4.1% of GNP respectively (OECD, 1988, p. 178).

The fact that the increase in the budget deficit was mainly a result of increasing government expenditure on transfers to households is important for the questions of the effect of government deficit on expenditure in the private sector. As this expenditure does not produce a direct government demand for goods and services but creates demand through its influence on disposable incomes in the private sector, the real effect on income building depends on the private sector's propensity for expenditure. However, the increase in the budget deficit influences private expenditure through its effects on financial development.

The financial consequences of the government deficit will depend on the means of financing it. If it is financed in the banks or in foreign markets it will result in an increase in liquidity in the banks and stimulate an increase in credits to the private sector. In such a case it will not result in rising interest rates. Its effects on private expenditure depend on factors such as capacity utilization in industry and income growth expectations. If the increasing budget deficit is financed on the bond market the effect on private expenditure is more uncertain. In this chapter we shall try to disregard the effects resulting from international relations. They will be treated in the next chapter. However, even in a closed economy the result of an increase in the financing of the government deficit on the bond market may have very different effects on private expenditure depending, for instance, on the actual level of capacity utilization and the production growth, on the structure and development of asset markets and so on.

We have examined earlier Benjamin Friedman's studies of the effects of a government deficit financed on the bond market. He concluded that the question of how private investment reacts to a bond-financed government deficit 'reduces to the long-debated issue whether bonds are closer portfolio substitutes for money or for capital' (Friedman, 1978, p. 618). Friedman summarizes the relevant substitutability properties in a relative substitutability index, defined as

$$\sigma \equiv \frac{m_2}{b_3}$$

where $m_2 = \partial M/\partial r_B$ is the marginal coefficient of money with reference to the bond rate and $b_3 = \partial B/\partial r_K$ is the marginal coefficient of bonds with reference to the yield of capital. If bonds are close substitutes for money but not for capital, m_2 is large and b_3 is small so that σ is large. If bonds are close substitutes for capital but not for money, m_2 is small and b_3 is large so that σ is small.

In Friedman's analysis there is an increase in government expenditure which leads to a rise in equilibrium income. The rise in equilibrium income is partly the result of the ordinary multiplier effect and partly the result of a wealth effect as can be seen from Friedman's equation for the equilibrium income Y (Friedman, 1978, p. 614):

$$Y = Y_0 + Y_1 G + (1 - Y_1)T + Y_2 r_K + Y_3(M + K + B)$$

The wealth effect will depend upon the wealth coefficient ratio

$$\frac{\partial M/\partial W}{\partial K/\partial W}$$

where W is wealth. Friedman denotes the wealth coefficient ratio by σ^*.

The influence of the relative substitutability index, i.e. the interest rate coefficient ratio, depends upon the value of σ relative to σ^*. If σ is smaller than σ^*, the increase in government purchases will cause an increase in r_K which will lead to a portfolio crowding-out effect. The financial disturbances caused by the fiscal policy expansion will reduce private investment, and the income expansion, which is a result of government expenditure, will be reduced. If, instead, σ is greater than σ^*, the increase in government expenditure will lead to a decrease in r_K and provoke an increase in private investment, while there will be an extra increase in the equilibrium income level. In this case there has been a portfolio crowding-in effect.

In Chapter 3 we studied the hypothesis that the interest coefficients are dependent on the growth rate of the economy. It is also probable that the wealth coefficients vary during the business cycle. Therefore whether fiscal policy measures result in crowding-out or crowding-in effects, as defined by Friedman, depends on the changes of both σ and σ^* during the business cycle.

An application of Friedman's theory to the fiscal policy in the industrial countries during the period of economic stagnation in the 1970s and 1980s should also take into account the fact that the large deficit in many countries was mainly an effect of increased transfers to the private sector and to a minor extent the result of increased expenditure on goods and services. This will give a smaller initial multiplier effect than in Friedman's example.

If the income increase is reduced it will result in a smaller increase in savings, i.e. in the increase in net wealth. The result would be less scope for additional investment in government bonds. This means that the investment capacity for a given issue of government bonds is less than it would have been if the budget deficit had been the result of increased expenditure on goods and services. Investment in government bonds will not result in a high increase in interest if investors are willing to increase the share of bonds in place of money but not at the expense of capital. In that case the value of the relative substitutability index would be high. It would also help if the wealth coefficient ratio was low, so that only a small part of the increase in wealth was accounted for by an increase in money.[3]

(a) *Structural changes in the credit system: some consequences for monetary policy*

A new trend in international financial development started in the mid-1970s when rising oil prices led to balance-of-payments deficits and increasing inflation in oil-importing countries, which included both industrial and developing countries. The deficit was financed by the oil-producing countries who exported the greater part of their oil revenues. The banking system in the industrial countries made this credit expansion possible by initiating an internationalization of the credit system which in the long run made it impossible to maintain the national regulation of the credit market.

Much of the balance-of-payments deficit in industrial and developing countries was financed by bank loans. Inflation and exchange rate variations increased the lending risks and led to rising bank rates and higher liquidity preferences. The banks had to expand their acquisition of funds. As they had to pay market interest rates their funding costs increased. They also encountered more competition in their deposit markets both from enterprises outside the banking system and from governments who adjusted their methods of borrowing to suit short-term investors. These new methods of government borrowing were also a consequence of the increasing budget deficit in many countries. They depended to a large extent on the fall in the growth rate, which led to a lower increase in tax income while government expenditure increased, particularly for transfers for industry subsidies and the unemployed.

The new situation which developed during the 1970s with a decreasing growth rate and increasing deficits in both the balance of payments and government budgets created new problems for the monetary policy. At the same time the internationalization of financial development made many of

[3]Werin (1981, p. 68), in his study of the Swedish budget deficit, stressed that the sectors have had stable portfolio distribution despite rather large changes in yields and market valuations. The sectors' portfolio choice is characterized by strong complementarity and very little substitution. Some important portfolio effects are as follows:

1. a substantial rise in the bond interest rate;
2. rising market values for equities and single-family houses;
3. a small increase in the market value for blocks of flats.

The concurrent increases in bond rates, equity values and the price of small houses confirm the complementary portfolio relations which Werin has commented on. The decrease in the value of blocks of flats is an effect of rising interest rates. The effect of the budget deficit has been partly stimulating and partly restrictive. The increase in equity prices amd the prices of small houses should be stimulating for private investment and lead to crowding-in effects. Conversely, an increase in the bond rate will keep investment down. Although Werin is not very sure about the net effect, he believes that the crowding-out effect has dominated as the rise in the bond interest rate was very large.

the regulations used in monetary policy during earlier decades less efficient.[4] Regulation as a means of imposing monetary policy became inefficient when it isolated the central bank from market-oriented credit flows. It became necessary for the central banks gradually to remove the regulations on credits and portfolio distributions in banks and other credit institutes as well as regulations on capital flows between countries. By the end of the 1980s monetary policy in most Western industrial countries was governed by the market.

(b) *An example of a model for monetary policy in the 1980s*

This model was published as a supplement to the medium-term prognosis for the Swedish economy for the period 1990–2000 (Englund *et al.*, 1987). Monetary policy in Sweden is now based on free-market adjustment. The last regulations disappeared in 1989 with the abolition of the foreign currency rules.

The effect of a monetary policy can be analysed in two steps. The first step is to obtain relations between the intermediary targets of the monetary policy such as the quantity of money, the flow of foreign exchange, interest rates etc. The second step is to obtain the relations between these intermediaries and the ultimate targets of economic policy such as production level, unemployment, balance of payments, price increases etc. Englund *et al.* adopt a Keynesian model in which the central bank, by controlling liquid assets and interest rates, will make the economy achieve given targets for expenditure and income.

The problem in a small economy which depends on international economic development is that interest rates must be tied to international interest rates if the central bank is to be able to control the capital movements towards the foreign market. This is necessary in order to keep exchange rates stable, which is the most important aim of central bank policy.

Englund *et al.* emphasize that the relations between the intermediary targets of monetary policy and the ultimate targets for the country's economic development cannot be taken as completely stable. They have also emphasized that in their expectations of future development the agents take into account the measures that they anticipate that the government will take. Therefore it may be impossible for the government to achieve the intended aims of its policy.

[4] An example can show how the development of business deposits with a market-adjusted interest rate in Sweden completely changed the banks' reactions to changes in the liquidity ratios, which was the central bank's most important means of controlling bank credits. An increase in this ratio used to make the banks decrease the growth of bank loans in order to increase the ratio of their liquid assets to their deposits. The creation of a market for short credit makes it possible for banks to increase their liquidity by borrowing on this market instead of reducing the growth rate for bank loans (Englund *et al.*, 1987).

The pessimistic view adopted by Englund *et al.* about the possibility of an efficient monetary policy is also revealed by their hypothesis that the relation between the demand for money and a few variables such as the real income and interest rates for bonds and foreign assets may only be stable in the long run. It is not a great help to the monetary policy of the central bank, which is concerned with short-run transactions in financial markets, that such long-run relations may exist. Englund *et al.* are also aware of the fact that short-run control by the central bank is important as it determines expectations of development in the medium and long term.

(c) *Model for monetary policy*

The model comprises four sectors: a public sector including households and all enterprises excluding banks, a bank sector, a government sector and a sector for the central bank. The assets and liabilities of these four sectors can be indicated in a transactions table (Fig. 6.8), which also includes a foreign sector. Even if the model is not concerned with economic events in foreign countries, there are foreign assets in some sectors and so transactions with the foreign sector must be possible.

Only the asset and liability stocks which are of interest for the monetary policy are included in the asset and liability matrices of the transactions table. The real capital as well as the funds representing long-term financial investments are absent. It is also important to note that the sector aggregation leads to the netting out of important financial stocks. As households, non-financial enterprises and all financial enterprises except banks (such as pension funds, insurance companies and other capital market

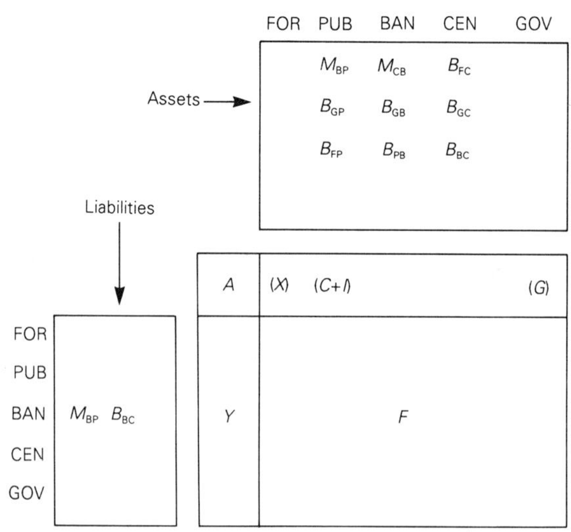

Fig. 6.8 Transactions table for a monetary policy model.

institutes) are included in the public sector, all equities, trade credits, concern claims and overnight loans between banks are netted out. The fact that equities and trade credits are not present in the model has to be interpreted as if either these assets are so sluggish in the short run that changes in them can be left out of account or they are such good substitutes for bonds that they can be included in this aggregate. Other assets have also been eliminated from the asset matrices because of the short-term character of the monetary policy. The time horizon is between a week and a quarter. In the asset matrices the public bank loans have been put in square brackets as they can be regarded as given in this short time perspective. As the interest rates for bank deposits are also given, bank deposits are determined by the public's demand for money.

The commercial banks' demand for reserves M_{CB} is determined by a reserve ratio imposed by the central bank. The reserve ratio a is the ratio of M_{CB} to the public's holdings M_{BP} of deposits in the commercial banks. Then

$$M_{\text{CB}} = aM_{\text{BP}} + \Sigma_t$$

where $a = M_{\text{CB}}/M_{\text{BP}}$ and Σ_t is a stochastic term for deviation.

As both bank loans and bank deposits are given in this short-term analysis there is just one portfolio choice open to the banks, i.e. the choice between investments in the short bonds included in B_{GB}, and loans B_{BC} in the central bank. Demand for such loans is assumed to be a function of the difference between the short-term bond rate r and the central bank discount rate d:

$$B_{\text{BC}} = b(r - d_1 X_b) + \Sigma_b$$

In addition to the stochastic term Σ_b the function includes the vector X_b of other explanatory variables. As well as representing the short bond rate, r can also refer to the interest rate for overnight loans between banks. Such loans are therefore regarded as perfect substitutes for short-term bonds.

The supply of non-borrowed reserve R can be defined as the total assets in the central bank excluding bank loans, i.e. the sum of the central bank's foreign assets B_{FC} and government assets B_{GC}:

$$R = B_{\text{FC}} + B_{\text{GC}}$$

The central bank's control of the non-borrowed reserves is managed by its open-market operations, i.e. by its purchases and sales of bonds on the market. With the help of this instrument the central bank can control the supply of liquid assets and thus the short-term market rate. However, it should be remembered that in a small open economy this market rate must be related to the international interest rate in order to obtain sufficient capital imports to keep foreign currency reserves stable. This is a condition for stable exchange rates.

The condition for equilibrium on the reserve markets can be written as

$$aM_{\mathrm{BP}} + \Sigma_t = R + b(r - d_1 X_b) + \Sigma_b$$

The portfolio behaviour of the public can be substantially simplified in the short perspective of monetary policy.

In this short-term analysis the public's bank loans are assumed to be given. We can also disregard the public's stock of banknotes and coin. The public's foreign assets have also been omitted from the model, although such assets could well be of sufficiently short-term character to be of interest in the analysis.[5]

The two remaining assets in the public's portfolio are bank deposits and bonds. These are assumed to be substitutes. The demand for money is related to the short-run interest rate r and to a vector X_m of other explanatory variables:

$$M \equiv M_{\mathrm{BP}} = m(r_1 X_m) + \Sigma_m \qquad \frac{\mathrm{d}M}{\mathrm{d}r} < 0$$

The equilibrium condition can be determined by substitution:

$$m(r_1 X_m) + \Sigma_m = \frac{1}{a}[R + b(r - d_1 X_b) + \Sigma_b - \Sigma_t]$$

The equation gives equilibrium values for the two possible targets of monetary policy r and M. The instrument for monetary policy is the value R of non-borrowed reserves. The central bank can also control the commercial banks' cash ratio a. In addition there are the two exogenous variables X_b and X_m and the three stochastic disturbance factors Σ_m, Σ_b and Σ_t.

It is interesting to compare Lybeck's analysis (Lybeck, 1975) of the regulated money market which existed in Sweden at the beginning of the 1970s with the situation at the end of the 1980s when the monetary policy was based on free-market adjustment. One fundamental aspect of the monetary policy has not changed. Sweden is still a small open economy. Stable exchange rates require a short-term interest rate which is adjusted to the international short-term interest level. In Lybeck's model the interest rate does not determine the demand for deposits. There are no financial substitutes for deposits in the non-bank private sector. Instead its potential to shifts into and out of real capital makes the expected rate of inflation the main opportunity cost. A quantity-theoretical approach can be logical with such a credit structure. A graphical illustration of the monetary market (Fig. 6.9) shows a horizontal supply curve at the given interest rate and a vertical demand curve as, according to Lybeck, the demand is independent of the

[5] This omission can be explained by the fact that Swedish foreign currency regulations were still in force when this study was published in 1987.

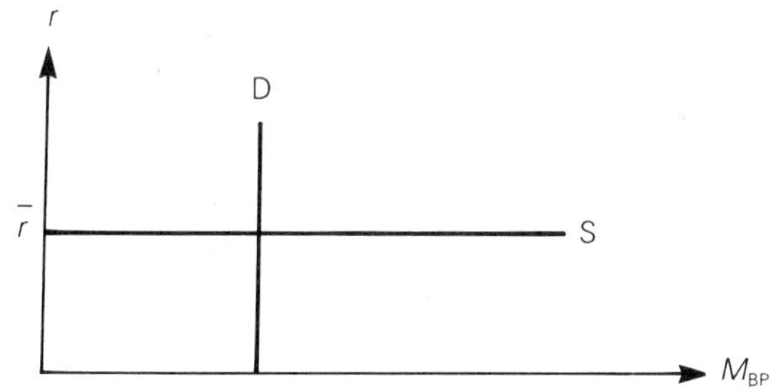

Fig. 6.9 Equilibrium when demand for deposits is independent of the rate of interest.

interest (Lybeck, 1975, p. 391).[6] The demand for deposits is completely inelastic and decides the amount of money. The supply is completely elastic at the given interest rate $\bar{r}$.

In the analysis by Englund *et al.* the interest rate is also given by the dependence on the international interest rate. As the supply of bank loans is also given in this short-term analysis the deposits are determined by the demand for deposits of this given interest rate. Unlike Lybeck, Englund *et al.* assume that the demand for money depends on the interest rate (Fig. 6.10).

An important structural change, also characteristic of the industrial world, has taken place in the Swedish asset market since the beginning of the 1970s. There are more types of credit institutes and there are portfolios with many more alternative approaches to financial investment for the public and for the banks. The banks also have a variety of avenues for borrowing and are no longer dependent on deposits for financial resources. This development of the credit market makes the Keynesian theory for demand of money a much more realistic component of a theory of monetary policy than it was at the time of Lybeck's analysis of the Swedish financial system.

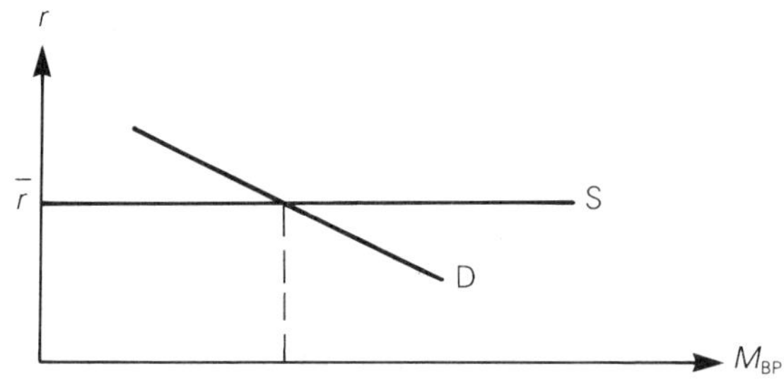

Fig. 6.10 Equilibrium when demand for deposits is elastic with respect to interest rate.

[6] In other parts of his book Lybeck has assumed the demand for money to be dependent upon the interest rate. But this summary in Chapter 10 shows his point of views more clearly.

However, Englund *et al.* appear to be sceptical about the appropriateness of a Keynesian approach. This doubt seems to depend upon the uncertainty regarding the relation between the central bank's objectives and the targets for income stabilization. It is also expected that monetary policy will have insignificant effects on real capital investment in a period of low capacity utilization and great uncertainty about future demand development—a fact that is also in agreement with Keynesian theory.

(d) *The operation of the Swedish model for monetary policy*

Englund *et al.* describe how the model can be used to simulate the process of monetary policy. We shall therefore define the intermediary targets of monetary policy, the instruments that the central bank can use and finally the exogenous disturbances which threaten to prevent the desired targets from being reached. The intermediary targets are the money stock M, the foreign currency reserve B_{FC}, interest rates r for overnight loans between banks and interest rates r_T with different maturity. The instruments are open market operations T^c, borrowing conditions $g(\cdot)$ in the central bank and reserve ratio a. Exogenous disturbances include foreign interest rate r_F, balance of trade CAB and other variables in the X vector, such as changes in domestic demand.

Figure 6.11 shows a model for the money market where the g curve indicates the supply of overnight loans under the assumption that $r = r_T$, so that there is only one endogenous interest rate. The stepwise form of the supply curve is a result of the central bank's rules about the quantity of credit that the commercial banks can receive at different levels of the central bank's special discount rate ('straffränta'). The demand curve shows how the banks' demand for reserves is a negative function of the interest rate. There are two reasons for this. If the domestic interest rate is high relative

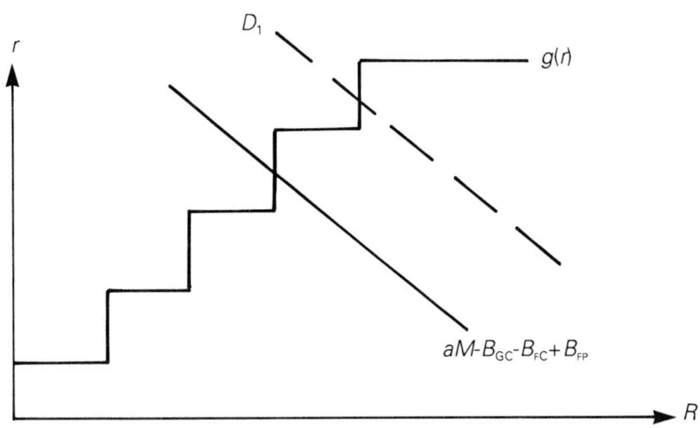

Fig. 6.11 The market for overnight loans in Sweden.

to the international rate, funds in foreign currency will be transferred to investments in domestic claims. This will result in increasing bank reserves, so that the need for bank borrowing from the central bank will decrease. The other reason is connected with the Keynesian liquidity preference. At a higher interest rate banks will expect that higher bond prices and lower market rates will be more probable in the future and will therefore increase investments in bonds and reduce their reserves.

Englund *et al.* (1987, p. 32) analyse the result of exogenous disturbances for the balance on the money market. In one case either there has been a change in the international economic development so that international interest has risen or the domestic demand for finished goods has increased in a way that may threaten the cost–price development. The central bank therefore wishes to increase the interest rate. The instrument for this is open-market operations which will tie up funds in bonds in the banks and in households, and so reduce the propensity to make expenditure on the goods markets.

The central bank's offer of bonds from its portfolio creates a supply surplus on the market and leads to falling bond prices and increases in the bond rate. The bank's need of reserves will increase, which leads to a shift in the demand curve to the right. The new equilibrium interest rate on the money market, which in this simplified model is equal to the equilibrium bond rate, is given by the new position of the demand curve.

In another case it is assumed that the exogenous disturbance is a deficit in the balance of payments so that the foreign currency reserves are reduced. The result will be a reduction in the total asset of the central bank and corresponds to a decrease in the reserves in the bank sector's assets and a reduction in importers' deposits in the bank sector's liabilities. In this case it will be the banks and not the central bank which will try to restore the normal level of reserves by selling bonds. The supply surplus leads to falling bond prices and a rising interest rate. The new equilibrium on the monetary market in Fig. 6.11 will indicate the new level of the interest rate.

REFERENCES

Englund, P., Hörngren, L., Viotti, S. and Vredin, A. (1987) Penningmarknad, räntebildning och valutaflöden, SOU 1987:3, Parliamentary Report.

Friedman, B. N. (1978) Crowding out or crowding in? Economic consequences of financing governments deficits. *Brookings Papers on Economic Activity*, 3.

Kaldor, N. (1982) *The scourge of monetarism*, Oxford University Press, Oxford.

Lundberg, E. (1963) *Penningpolitikens Brister och Möjligheter*, Skandinaviska Bankens Kvartalsskrift, Boktryckeriet, Stockholm.

Lybeck, J. A. (1975) *A Disequilibrium Model of the Swedish Financial Sector*. Libertryck, Stockholm.

Lybeck, J. A. (1978) *Konjunkturer och Svensk Ekonomisk Politik*, Esselte Herzogs Combiflex Nacka, Stockholm.

OECD (1988) *Economic Outlook*, **44**.

Pedone, A. (1988) Public expenditure. In A. Boltho (ed.), *The European Economy: Growth and Crises*, Oxford University Press, Oxford.

Radcliffe Report (1959) *Committee on the Working of the Monetary System*, HMSO, London, Cmnd 827.

Werin, L. (1981) *Perspektiv på Budgetunderskottet*, Del 3, DsFi 1953:29.

Werin, L. (1981) Budgetunderskott, Portföljval och Tillgångsmarknader in *Perspectiv på Budgetunderskottet*, DsFi 1983:29 (Report from the Ministry of Finance) Liber, Stockholm.

7

Financial stability and growth in an internationalized economy

7.1 STRUCTURAL CHANGES IN PRODUCTION AND THE DEPENDENCE ON FOREIGN TRADE

(a) *Introduction*

In most industrial countries the influence of financial targets and constraints on the level of internal spending is channelled through changes in foreign trade. Changes in internal spending generally feed back into external trade flows, particularly in the form of demand for imports. They also have effects on export development through their influence on the relation between foreign and domestic price development and on the production capacity in the export sectors.

Changes of imports and exports have an immediate effect on the financial development as a result of changes in the foreign currency reserves, which alter the liquidity position of the banking system. Cyclical development in many industrial countries, particularly in Western Europe, is dominated by the growth of exports which often is not synchronized with the growth of imports. The result will be a cyclical pattern in the changes in foreign currency reserves and liquidity in the banking system.

The Bretton Woods reforms at the end of the Second World War created a system with stable exchange rates and institutions such as the International Monetary Fund and the World Bank, which were meant to construct an international system of credits sufficient to cover short-run cyclical disequilibria in the balance of trade. It was soon apparent that the destruction of the production capacity in many industrial countries during the war made a more comprehensive adjustment necessary. The reconstruction with the help of Marshall Aid in Western Europe during the period 1947–52 did much to re-establish the production capacity. However, it was also necessary to make repeated adjustments in the exchange rates in almost all Western European countries. The result was devaluation relative to the US dollar of all European currencies, except the deutschmark and the Swiss franc, during the period of high growth rates in the 1950s and 1960s. The adjustments of the exchange rates led to financial upheavals which outran the financial changes resulting from the normal cyclical variations in factor income distribution. The financial disturbances during the period of high growth rate indicate that there was incompatibility between the structural changes in these countries and the high growth rate of their spending targets. However, this conflict did not lead to a collapse until the next decade when the Bretton Wood system was abandoned. The oil price crises resulted in a high world-wide inflation, which was not brought under control until the beginning of the 1980s after some years of very restrictive policy. This policy resulted in an increase in unemployment which reached pre-war levels in some countries and had not returned to normal levels at the end of the 1980s.

In order to clarify how international relations influence the interrelations between structural change and financial development it is necessary to compare economic development in different groups of countries. We shall start with a study of relations between trade policy and productivity development (Haaland *et al.*, 1986).

(b) *Trade policy and productivity*

Haaland *et al.* (1986) assume that a liberalization of trade will lead to improved resource allocation. He starts from the international situation in 1979 when a system of protective trade rules existed for five commodity groups in six regions. The commodity groups were knowledge-intensive goods, labour-intensive goods, capital-intensive goods, energy and raw materials. The regions were Western Europe, North America, Japan, newly industrialized countries (NICs), countries belonging to the Organization of Petroleum Exporting Countries (OPEC) and the non-oil developing countries (NODCs).

(c) *The stand-still scenario*

The approximate productive effect of trade and industrial policy on the five commodity groups in the six regions in 1979 was evaluated as a percentage of value added and is shown in Table 7.1. The stand-still scenario means that the protection system described above will remain unchanged in the 1990s.

The data give the combined effect of both tariff and non-tariff measures. Liberalization on both a global and a regional basis has reduced the importance of customs duties since the Second World War. However, non-tariff measures have increased since 1970. Table 7.1 shows that the protective level—weighted according to the share of the commodity in regional production— is about twice as high in Western Europe as in North

Table 7.1 The protective effect of trade and industrial policy in 1979 (percentage of value added)

	Western Europe	*North America*	*Japan*	*NICs*	*OPEC countries*	*NODCs*
Knowledge-intensive industrial products	15	10	10	25	25	40
Labour-intensive products	45	30	15	10	25	20
Capital-intensive products	30	18	12	25	25	40
Fuel	30	15	26	16	0	10
Raw materials	124	10	130	68	61	10
Total	52	16	34	32	14	21

Source: Wergeland 1985.

America, the OPEC countries and the NODCs. Japan and the NICs are about midway between. The high protective level in Western Europe depends upon the fact that non-tariff protection is included and that production shares and not the countries' own imports are used as weights. When imports are used as weights the protection is underestimated, while production shares as weights give an overestimation.

Western Europe and Japan protect their raw materials, including agricultural products, three to four times as much as the countries in the other regions. OPEC countries and NICs protect raw materials about twice as much as industrial products. North America and the NODCs protect industrial products more than raw materials. North America and Western Europe protect labour-intensive products about twice as much as other industrial products. The NICs and the NODCs, however, protect capital-intensive and knowledge-intensive industrial products twice as much as labour-intensive products.

(d) *The roll-back scenario*

Liberalization is assumed to involve equivalent concessions according to the reciprocity principle used in negotiations under the auspices of the General Agreement on Tariffs and Trade (GATT). Customs concessions are extended to other countries according to the most favoured nation clause. Three groups of agreements are particularly important: reintroduction of trade in agricultural products according to GATT rules, graduation of NICs to full GATT status and regulation of the trade between Japan and the European Community (EC) according to GATT principles.

Liberalization is seen as an abolition of the trade barriers which have gradually developed.

Subsidization of agricultural products in the EC and the United States has led to high costs for both government and society in the form of an unproductive resource allocation. It has caused difficulties in countries with a large volume of agricultural production such as Australia, Canada, Thailand and Argentina. A decrease in agricultural subsidies in Western Europe and Japan could be compensated by abolition of restrictions on trade in knowledge-intensive goods in the United States.

The NICs have had advantages, as developing countries, over the old industrial countries. However, they are interested in entering markets in the traditional industrial countries where reciprocal liberation may be possible.

Both Western Europe and the United States want to open up domestic markets in Japan, although the disequilibrium in trade between the United States and Japan may be a result of macroeconomic policy rather than trade policy restrictions.

(e) *The roll-on scenario*

The third scenario is characterized by a continuation of the existing trade restrictions during the 1990s. High unemployment in Western Europe and increasing competition from Japan and the NICs will lead to an increased protectionism in Western Europe. In order to maintain market shares there will be increased protection of agricultural, labour-intensive and knowledge-intensive products in Western Europe. There will be increased protection for knowledge-intensive products in all regions, and greater protection of capital-intensive products in the developing countries.

The result will be increased structural rigidity. Europe will continue to protect both its low-technology and its high-technology products.

(f) *The consequences of trade policy development for growth and productivity in the six regions*

The yearly growth rate of the labour force and production have been calculated for the three scenarios for the period 1979–95. Therefore it is possible to compare the growth rate in productivity measured as the difference between growth of production and growth of the labour force.

Table 7.2 shows that the roll-back scenario with trade liberation would be most advantageous in Western Europe, Japan and the NICs. The GNP growth rate in Western Europe would be 2.5% and the productivity growth rate would be 2.05%. In Japan the GNP growth rate would be nearly double the European level at 4.8%, and the productivity growth rate would be 4.1%. Even the NICs would benefit from trade liberalization. The roll-back scenario gives them a GNP growth rate of 4.3% and a productivity growth rate of 1.55%.

For the other regions the stand-still scenario gives the highest growth

Table 7.2 The effects of three alternative trade policy scenarios on GNP and productivity growth rates in six regions

	Growth rate						
		Roll-on scenario		*Stand-still scenario*		*Roll-back scenario*	
	Labour	*Productivity*	*GNP*	*Productivity*	*GNP*	*Productivity*	*GNP*
Western Europe	0.45	1.45	1.9	1.65	2.1	2.05	2.5
North America	1.10	1.9	3.0	2.0	3.1	1.9	3.0
Japan	0.70	3.9	4.6	3.8	4.5	4.1	4.8
OPEC countries	2.70	0.5	3.2	0.6	3.3	0.5	3.2
NICs	2.75	0.75	3.5	1.45	4.2	1.55	4.3
NODCs	2.50	0.3	2.8	0.8	3.3	0.8	3.3
Total	—	—	2.9	—	3.1	−3.2	—

rates for GNP and productivity, except for the NODCs which will have the same growth rates in the roll-back and in the stand-still scenarios, i.e. 3.3% for GNP and 0.8% for productivity. The OPEC countries will have the same growth rate for GNP (3.3%), but only 0.6% for productivity. North America will have about the same productivity growth rate in the stand-still scenario (2.0%) as Western Europe (2.05%) in the roll-back scenario, but a higher GNP growth rate (3.1%) as the labour growth rate is 1.1%, more than twice the rate in Western Europe.

The conclusion of this study is that, in Western Europe in particular, liberalization of trade barriers and abolition of both tariff and non-tariff measures would result in a substantial increase in productivity.

Assumptions about factors explaining the potential production growth rate in the six regions in the period 1979–95 are also presented in the study. Table 7.3 shows the growth rate of the labour force divided into highly educated and poorly educated workers. It is also possible to compare capital building by using the data for net saving. Productivity has been divided in two series—one for 'old capital' and one for 'new capital'. The productivity growth rate is higher for new capital than for old capital. However, the same growth rate has been indicated for all country groups. Haaland (1985, p. 18) admits that this result is not realistic; it is only a starting point for further analysis.

Some explanation of the differences found in productivity growth in comparing production and labour force growth in the three trade policy scenarios can be found in the data concerning the relation between the growth rate for the highly educated and the poorly educated labour force in the six regions as well as for the different levels of the net savings ratio. Differences between the growth rates of the highly educated and the poorly

Table 7.3 Assumed yearly growth rates for supply factors in six regions, 1979–1995 (%)

	Western Europe	*North America*	*Japan*	*OPEC*	*NIC*	*NODC*
Labour						
highly educated	1.25	1.90	1.50	2.70	2.75	2.50
poorly educated	0.35	1.00	0.60	2.70	2.75	2.50
Supply of low-cost energy	—	1.95	—	1.95	—	1.95
Capital depreciation	−3.50	−3.50	−3.50	−3.50	−3.50	−3.50
Net saving	10.00	7.50	15.00	7.50	7.50	7.50
Capital import	−2.50	−2.50	0	0	27.40	0
Productivity						
'Old' capital	0.50	0.50	0.50	0.50	0.50	0.50
'New' capital	1.50	1.50	1.50	1.50	1.50	1.50

educated labour force are found in the three industrial regions (Western Europe, North America and Japan), whereas in the three regions containing developing countries (OPEC, NICs and NODCs) the two growth rates are the same.

The high net savings ratio in Japan will facilitate a rapid increase in capital-intensive production after liberalization of the trade policy. Western Europe and North America will increase their shares in knowledge-intensive production, although the increase in North America will be greater in the roll-on scenario than in the roll-back scenario.

(g) *Financial growth and external disequilibrium*

The dominant international economic problem since the 1970s has been the financing of the deficits in the US economy. The balance-of-payments deficit developed during the period of sustained growth before 1973 and led to the dollar devaluation in 1971. This devaluation contributed to the subsequent price increase in oil and other raw materials.

Despite rising prices and continuing deficits in both the balance of payments and the government budget US policy has continued along an expansive trend. This has been made possible by the US economy's dominant international position. Its high import growth rate has led to high export and production growth in other industrial countries. As a financial centre the United States has been able to obtain the capital import necessary to finance its deficits. Its dominance of the international equity and bond markets has allowed US authorities to adjust bond prices in such a way that savings in foreign countries were invested in the United States. It was also vital that the bond rate did not become so high that the industrial growth in other countries was threatened or growth in equity prices obstructed. These adjustments of interest rates have led to changes in the dollar value which have influenced both the terms of trade and the international competition of US industry.

In order to clarify how economic development in the United States has influenced the world economy through its effects on the international financial situation we use a small model.[1]

The model refers to the medium-term development and has only three sectors—the foreign sector, the private sector and the government sector. The symbols assigned to the variables refer to changes compared with a given initial position. Capital roman letters indicate changes in nominal values, small roman letters indicate quantitative changes and greek characters indicate coefficients.

[1]The model used here is based on a model given by Tinbergen (1966, p. 249) and analysed by Kragh (1967, pp. 38–50). It has been modified in order to include more of the financial development.

The national income is given by

$$Y = X + E - I \tag{7.1}$$

where X is home market expenditure, E is exports and I is imports. Home market expenditure is given by

$$X = X_0 + \xi Y \tag{7.2}$$

where X_0 is the autonomous part, which is assumed to refer to government expenditure and whose level is ruled by political targets. The derived part of home market expenditure includes both investment and consumption. It has been treated in a very schematic way as a function of the change in nominal national income. This relation must be assumed to cover both the dependence of private consumption on disposable income and household wealth and the dependence of private investment on capacity utilization and Tobin's q. Such an approach may be valid in a medium-term analysis.

The change in export volume is dependent on the international product growth rate, which is dominated by US import expenditure i:

$$e = \varepsilon_1 i - \varepsilon_2 r - (p^e - p^w + k) \tag{7.3}$$

The international growth rate is also dependent upon the financial development. An increase in the interest rate r will keep down the growth in international demand and result in lower growth in US exports. US exports are also dependent upon a change in the relation between export prices p^e and the international price level p^w. k indicates the effect of a change in the exchange rate. An appreciation gives a positive value for k and is equivalent to a rise in the export price for foreign countries. Changes in prices can be expressed as a percentage of the prices of the base period. The price of the base period is denoted $\bar{p}^e = \bar{p}^w = 1$. The two price-fixing equations

$$p^x = \pi_1^x v + \pi_2^x (p^i - k) \tag{7.4}$$

$$p^e = \pi_1^e v + \pi_2^e (p^i - k) + \pi_3^e (p^w - k) \tag{7.5}$$

show how changes p^x and p^e in domestic and export prices are determined by changes v and p^i in the production volume and the import price with regard to changes k in the exchange rate. The coefficient π_1 shows how prices react to variations in capacity utilization. It is influenced by both the Verdoorn effect and the Phillips effect. The coefficient π_2 shows the share of imports in production costs. The export price has also been taken into consideration in the rate of changes in world-market prices and changes k in the exchange rate.

Changes in the exchange rate are a function of changes in the nominal value of the balance of goods and services:

$$k = KD \tag{7.6}$$

Changes i in the import volume depend on changes v in the total production, which is the sum of home market production x and exports e:

$$i = \alpha v \tag{7.7}$$

$$v = x + e \tag{7.8}$$

The three equations

$$\mathrm{X} = xp^{\mathrm{x}} + \bar{x} \tag{7.9}$$

$$\mathrm{E} = ep^{\mathrm{e}} + \bar{e} \tag{7.10}$$

$$\mathrm{I} = ip^{\mathrm{i}} + \bar{i} \tag{7.11}$$

define the changes X, E and I in the nominal value of home market production, exports and imports in terms of the changes xp^{x}, ep^{e} and ip^{i} of the prices of home products, exports and imports and the changes $\bar{x}$, $\bar{e}$ and $\bar{i}$ in volume of home production, exports and imports. The changes in prices are weighted by the quantities of the base year ($\bar{x}$, $\bar{e}$, $\bar{i}$).

Changes in the trade balance of goods and services in nominal values are a function of changes r in the interest rate, which is assumed to be controlled by the authorities, and the expected return on equities A^{p}:

$$D = I - E \tag{7.12}$$

$$D = \Psi_1 r + \Psi_2 A^{\mathrm{p}} \tag{7.13}$$

$$A^{\mathrm{p}} = \zeta_1 Y - \zeta_2 r - \zeta_3 t \tag{7.14}$$

Changes in the expected return on equities are assumed to be a function of changes Y, r and t in the nominal national income, the interest rate and the tax rate, which is an instrument of the financial policy.

The model can be used to study some of the international economic problems which have resulted from economic developments in the United States. The first question concerns inflation in the United States which was the main reason for the restrictive economic policy which was imposed after the second oil price crisis.

As the model includes 14 equations and has 14 endogenous variables we can eliminate all variables except the changes in the home market price level p^{x}. The expression for p^{x} includes the exogenous variables. Among these variables are those which can be controlled by the political authorities. They are changes r, X_0 and t in the interest rates, in government expenditures for goods and services, and in tax rates. Changes p^{w} and p^{i} in world prices and import prices cannot be controlled. The interest rate was the most important instrument used by the authorities to fight inflation after the second oil price crisis. The general solution is as follows:

$$p^{\mathrm{x}} = a_1 r + a_2 X_0 + a_3 t + a_4 p^{\mathrm{w}} + a_5 p^{\mathrm{i}}$$

The values of the coefficients depend upon the values of the coefficients chosen in the model. Kragh (1967, p. 42) chose values which, according to Tinbergen (1966, p. 250), apply to a small open economy like the Netherlands:[2] $\xi = 0.9$, $\Sigma = 1$, $\pi_1^x = 0.1$, $\pi_2^x = 0.3$, $\pi_1^e = 0.1$, $x = 1$, $\bar{i} = \bar{e} = 0.5$, $\pi_1^e = 0.2$, $\pi_3^x = 0.5$, $\alpha = 0.33$.

How would the values of the coefficients in our model applied to the US economy compare with those employed by Kragh and Tinbergen for a small economy?

The marginal propensity to spend ξ would probably be unchanged in a dominant economy. However, our export function has a different structure because of the dominant role of the US economy. Σ_1 is the elasticity of exports with respect to imports. Its value depends on the dependence of other countries on US demand. It should be quite important, particularly in the Western hemisphere. The same is true of Σ_2 which is the elasticity of exports with respect to changes in the US interest rate. In Latin America in particular the rise in the US interest rate has led to financial difficulties which have restrained import demand. These two factors are probably more important than the third factor—changes in the ratio of export prices to world prices. This is because of the dominant role of US price changes in world price development.

The greatest difference between the coefficient values in a model for the United States and for small open economies concerns the dependence upon imports. Even if this dependence has increased in the United States it is much smaller than in the open economies in Western Europe for instance. This results in a lower value for the import elasticity α in the model for the USA than in that for the small open economies. It also gives lower values for the coefficients π_2 in the two price-fixing equations which indicate the import share of production costs.

In conclusion, this comparison seems to indicate how it is possible for the US authorities, because of the influence of the United States on international economic development, to militate against the inflationary tendencies due to deteriorating terms of trade and to carry through an expansionary policy despite large disequilibria.

The model can also be used to illustrate the change in US policy which began in 1982 when international protests against the high interest rates

[2]It should be noted that the structure of Kragh's model is different from that of our model as it is assumed to be adapted for small open economies. For this reason the export function is constructed differently from that in our model with exports a function of only the relation between export prices and world prices with respect to changes in the exchange rate. It is written

$$e = -\varepsilon(p^e - p^w + k)$$

The model does not include the relations with financial development: r and the demand for equities are not included. The variables controlled by the authorities are the changes X_0 and k in the government's expenditure on goods and services and in the exchange rate.

forced the US authorities to employ other instruments to stimulate foreign capital imports in order to finance US deficits. Reduction in income taxes together with an increase in production growth rate led to an increasing demand on asset markets, particularly for equities. In the equation derived below we have eliminated all variables except the balance-of-payments deficit. In the expression a constant deficit can be obtained as the decreases in the interest rate and the tax rate lead to increasing production changes and to increasing equity prices. The expression for the balance-of-payments deficit is given by Equations (7.12)–(7.14):

$$D = I - E \tag{7.12}$$

$$D = \Psi_1 r + \Psi_2 A^{\mathrm{p}} \tag{7.13}$$

$$A^{\mathrm{p}} = \zeta_1 Y - \zeta_2 r - \zeta_3 t \tag{7.14}$$

On substituting we obtain

$$I - E = \Psi_1 r + \Psi_2(\zeta_1 Y - \zeta_2 r - \zeta_3 t)$$

As $Y = X + E - I$ this expression can be written

$$I - E = \Psi_1 r + \Psi_2 \zeta_1 [X - (I - E)] - \Psi_2 \lambda_2 r - \Psi_2 \zeta_3 t$$

$$(I - E)(1 + \Psi_2 \zeta_1) = (\Psi_1 - \Psi_2 \zeta_2) r + \Psi_2 \zeta_1 X - \Psi_2 \zeta_3 t$$

$$I - E = \frac{1}{1 + \Psi_2 \zeta_1} (\Psi_1 - \Psi_2 \zeta_2) r + \Psi_2 \zeta_1 X - \Psi_2 \zeta_3 t$$

The coefficients are as follows:

Ψ_1 is the elasticity of capital imports with reference to the interest rate
Ψ_2 is the elasticity of capital imports with reference to the expected return on equities
ξ_1 is the elasticity of equity demand with reference to nominal production changes and is assumed to be equal to 1
ξ_2 is the elasticity of equity demand with reference to changes in the interest rate (the absolute value of ξ_2 is less than 1 in growth and greater than or equal to 1 when growth is decreasing)
ξ_3 is the elasticity of equity demand with reference to the tax rate and is assumed to be zero (it can have different signs depending on the budget deficit and its influence on interest rate expectations)

(i) The effect of the interest rate on capital imports. We have

$$\frac{\delta(I - E)}{\delta r} = \frac{\Psi_1 - \Psi_2 \xi_2}{1 + \Psi_2 \xi_1}$$

As $\xi_2 < 0$, $\Psi_2 > 0$ and $\Psi_1 > 0$ an increase in the interest rate will always have a positive effect on capital imports. However, when there is growth the effect will be smaller as $|\xi_2| < 1$, i.e. the increase in the interest rate will reduce the demand for equities. When there is no growth the increase in the interest rate will have a strong positive effect on capital imports as $|\xi_2| > 1$.

(ii) The effect of growth on capital imports We have

$$\frac{\delta(I - E)}{\delta X} = -\frac{\Psi_2 \xi_2}{1 + \Psi_2 \xi_1}$$

The effect will be positive when $\xi_2 > 0$ as growth leads to foreign demand for equities.

(iii) The tax effect. We have

$$\frac{\delta(I - E)}{\delta t} = \frac{\Psi_2 \xi_3}{1 + \Psi_2 \xi_1}$$

The effect depends on the sign of ξ_3. Changes in the tax rate may influence expectations of production growth if, for instance, it is assumed that a decrease in the tax rate will increase future profits and lead to a higher investment growth rate. In such a case there may be a positive effect on equity demand. However, there could be a negative effect if it is assumed that the decrease in the tax rate will lead to lower government income and increasing budget deficit. In such a case the result may be expectations of a higher interest rate in the future which will keep down demand for equities.

Despite its simple structure the model can help to clarify how the deterioration in terms of trade produced inflation in the United States, and how this inflation forced the US authorities to a restrictive policy and the acceptance of a lower growth rate. However, it cannot show how the change in terms of trade led to a profound alteration in international income distribution, which had a dramatic effect on the position of the non-oil-producing industrial countries relative to that of the oil-producing countries. The domestic income distribution was also affected by inflation and by the efforts of various income groups to maintain their real income position. The period of stagnation from 1980 to 1982 led to substantial reductions in the profit share. In 1982 the US authorities had to reduce interest rates because of the grave financial problems that the high rates had produced in many countries. However, sufficient capital imports to finance foreign trade and budget deficits would be compatible with a lower interest rate, if the growth rate increased and the fiscal policy was adjusted to stimulate profit growth. In the years following 1982 the profit share increased and, together with increasing capital imports, led to a very high growth rate in equity prices.

(h) *Changing terms of trade in a small open economy*

The model used to analyse the role of international economic relations in the United States can also be used for a small open economy after changing the export function and the coefficients:

$$e = \Sigma_1 v^{w} - (p^{e} - p^{w} + k)$$

Thus in a small open economy the change in the export volume depends upon the change v^{w} in world production and a change in the relations between the country's export prices p^{e} and the international price level p^{w}. k indicates the effect of a change in the exchange rate. A devaluation gives a negative value for k and is equivalent to a fall in the prices which the foreign countries have to pay the small open economies for their export goods.

Solving the total model for changes in domestic prices with reference to the exogenous variables gives

$$p^{*} = a_1 r + a_2 X_0 + a_3 t + a_4 v^{w} + a_5 p^{w} + a_6 p^{i}$$

The development of the domestic price level in the small open economy will depend much more strongly on exogenous variables such as world production and world prices than is the case for the dominant economy. In those countries which were dependent on oil imports the first and second oil price crises hit the economy both by creating a balance-of-payments deficit and by increasing the costs for energy and other intermediary goods and services. When the balance-of-payments crisis led to devaluations the terms of trade deteriorated and the cost expansion continued. Inflation was increased further by the efforts of various income groups to maintain their real income position. The interrelation between wage and price increases has been analysed in earlier chapters. The process is important in this context as an explanation of the authorities' unwillingness to risk a situation where devaluation would be inevitable.[3]

(i) *Financial expansion and crises*

The oil price crises led to an income expansion in the OPEC countries. However, as the production capacity in most of these countries did not allow an expansion of domestic expenditure of equal proportions, most of

[3] Changes in the terms of trade also have a marked effect on cyclical variations in factor income distribution in small open economies. The start of cyclical expansion in such countries is often the result of increasing demand on the export markets. The growth rate of import prices may be low in this phase as a result of the low capacity utilization. Better terms of trade in the expansion phase will strengthen the increase in profit shares. When the recession sets in and the high growth rate in wages leads to falling profit shares this development may be strengthened by a stronger fall in export prices than in import prices. However, this development may vary depending on the product composition on the export and import sides.

the increase in oil income during the 1970s was saved. The savings were invested in the credit system of the industrial countries and from this they were channelled into financing the increasing balance-of-payments deficits of the industrial countries and of those developing countries who were not oil producers (NOU countries).

Although the production growth rate decreased in the industrial countries it was maintained in the NOU countries. This unexpected growth of liability financing of balance-of-payments deficits in the 1970s was not constrained by either the countries concerned or the banking system. It was demand determined.

The expansion changed the structure of the banks in the credit system. International interbank relations increased. In the 1970s banks did not act merely as intermediaries in placing bond issues from NOU countries on the market, but took direct exposure with a corresponding risk. This forced the banks to increase their risk premiums and enforce liquidity preferences. The increasing inflation expectation led to higher nominal interest rates.

An important part of the banks' asset and liabilities management was to determine ways of limiting the various types of risk which could be classified as interest rate risks, credit risk, foreign currency risks, liquidity risks and risks of losing shares on markets for financial services (Van der Hoeven, 1987, p. 39). One way of limiting credit risks was to reduce loans with fixed interest rate and instead adjust the interest on lending to the interest rates on deposits. A new policy was introduced according to which funds were no longer acquired in order to finance lending. Instead the lending conditions were adjusted to the conditions for borrowing. This was the result of a very important change in the banks' procurement of funds which had taken place. Instead of the passive reception of funds which the public deposited in different types of accounts, the banks had been forced by the competition for funds to borrow in different forms, for example, on the Eurodollar market, by issuing bank certificates or selling out of the diversified portfolios which became disposable during the 1970s.

However, this development on the credit market—the so-called 'securitization'—can also imply increasing risks for the financial system. The banks' increasing dependence on each other makes them very sensitive to even small disturbances on the markets. If an institute got into difficulties because of lack of liquidity and had to sell out of its portfolio, asset price changes could be produced which led to new problems in an increasing number of institutes. We shall see that this type of situation became common during the 1980s.

The second oil price shock, the restrictive monetary policy which followed and the recession in the industrial countries created serious problems in the NODCs. The financial development deteriorated with rising interest rates, and in 1982 the Mexican collapse occurred which marked the onset of the debt crises. Higher nominal rates also reduced liquidity by shifting the burden of debt repayments towards the current period. The combination of rising interest rates and increasing dollar currency prices for interest payments, which in this way came to include an increasing part of the NODCs' export incomes, was even more serious. Voluntary lending to the NODCs by commercial banks decreased after the Mexican crisis in August 1982. It has been said that it was a shift to credit rationing, rather than a continuous tightening of terms and constraints for borrowers (Eichengreen and Parkers, 1987).

The external shock in some developing countries is summarized in Table 7.4 (Sachs, 1985). The real income effect of the change in terms of trade is shown as the percentage change in terms of trade in the period 1979–83 compared with the period 1975–8 multiplied by the change in the import share as a percentage of GDP between the same periods. The weighted average shows a positive change, but there are very different developments in individual countries from a positive effect of more than 15% in Venezuela to a negative effect of almost 5% in Brazil. The interest rate shock is measured by multiplying the change in real interest rates between 1975–8 and 1979–83 by a debt-to-GDP ratio. Finally the two measures are added in order to indicate the role of external shocks in the developing

Table 7.4 Terms of trade shock, 1979–1983[a]

Country	*Percentage change in terms of trade 1975–8*[a]	*Real interest rate effect as percentage of GDP*[b]	*Sum of real interest rate and terms of trade effect*
Latin America			
Argentina	3	0.0(−0.3)	1.6
Brazil	−29	−2.7(−2.9)	−5.0
Chile	−27	−1.3(−1.9)	−6.2
Mexico	26	−0.7(−0.7)	1.2
Peru	−22	−0.5(−0.9)	−4.2
Venezuela	64	0.2(0.8)	16.2
Weighted average	1.1	−1.3(−1.4)	−0.1
Colombia	−18	−0.9(−0.9)	−2.8
East Asia			
Indonesia	36	0.1	6.2
Korea	−3	−2.9	−3.8
Malaysia	14	−0.1	4.8
Thailand	−14	−0.4	−3.3
Weighted average	11.8	−1.0	1.1
Philippines	−16	−0.7	−3.9

[a] Percentage change in average terms of trade for 1979–83 relative to average for 1975–8.
[b] To state the interest rate change as a fraction of GDP $r_{79-83} - r_{76-78}$ is multiplied by the ratio of net debt to GDP for 1980, where the net debt is the total debt net of foreign exchange reserves converted into the local currency at the official exchange rate.

countries. It should be noted that the countries which are particularly hard hit by the terms of trade effect, such as Brazil, Chile, Peru, Korea, Thailand and the Phillippines, have also suffered from the real interest rate effect.

7.2 CYCLICAL CHANGE AND FINANCIAL DISEQUILIBRIUM

(a) *The 'stop–go' problem*

It has been emphasized earlier that the long period of high growth rate in the industrial countries during the 1950s and 1960s was characterized by cyclical variations around a rising trend line. The fluctuation in employment and the inflation rate may seem unimportant in comparison with developments during the two following decades, with the exception of the inflation caused by the outbreak of the Korean War and the recession in the United States in 1957–8. However, the cyclical fluctuations produced temporary deficits in the balance of payments in some of the major industrial countries which had significant effects on financial development.

Since the Second World War the aim of financial policy in the industrial countries had been to counteract the cyclical variations. Capacity utilization in Western Europe remained at a high level during the 1950s and the 1960s. Even during the strongest recession the GNP gap was less than 4%. Therefore business fluctuations moved very near the 'ceiling' of full capacity. However, the high average capacity utilization during this period meant that there was shortage of skilled labour and raw materials at the peaks of the cycles. Imports expanded and trade deficits developed. These were reinforced when the high demand on domestic markets prevented export growth.

Even if these deficits were not very important in the major industrial countries and in the long run tended to equalize, they were of great importance in the short run. In the United Kingdom the existence of the sterling area and sterling's status as an international trading currency both exacerbated cyclical balance-of-payments problems and at the same time restricted exchange rate policy by making the maintenance of fixed parity a central aim of economic policy. As early as 1951, when inflation increased as a result of the outbreak of the Korean War, the balance-of-payments deficit led to heavy speculation against sterling. Import controls and hire purchase restrictions in order to keep consumption growth down were unsuccessful, and in 1952 the bank rate was revised for the first time since the war. After that the bank rate and the short interest rates were adjusted to the business cycle variations in order to influence international capital flows in such a way that speculation in sterling devaluation could be neutralized. Any expectation of devaluation was likely to lead to large outflows of short-term capital. This meant that any deterioration in the current account, however insignificant, led to suspicion of devaluation and unsustainable falls in the reserves. Economic policy became increasingly directed towards maintenance of the external equilibrium (Surrey, 1982). This became impossible in 1967 and sterling was devalued.

(b) *Exchange rate speculation in the transactions table*

The importance of speculation on foreign exchange markets increased with the liberalization of international trade and payments during the 1960s. Perhaps the most important financial innovation in this period was the creation of the Eurodollar market, which had a great importance for financing US investment with European funds. Financial investments in foreign assets can be illustrated in the transactions table. In Fig. 7.1 there are five sectors: one foreign sector (FOR), one private sector excluding credit institutes (PRI), one government sector (GOV) and two financial sectors, banks (BAN) and other credit institutes (OCI).

The portfolios of the private sector and the bank sector are essential for our study of the role of the foreign assets. As long as the growth rate is stable and the export revenues develop according to expectations the foreign currencies in the commercial banks are transferred to the currency reserves B_{FB} in the central bank. The growth of these reserves ensures a stable development of liquidity in the banking system and bank credit will be abundant. Interest rates will stay at a low level. When capacity utilization has reached such a high level that bottlenecks appear in the domestic economy the growth rate of imports will increase. Increasing domestic costs may also impair export rates. There will be a decrease in the growth rate of the currency reserves and in the banks' liquidity position. When production growth rates decrease income expectations deteriorate and asset values will fall. The risk of a devaluation will make exporters keep their revenues in foreign assets. There will be an increasing demand for short liquid foreign assets, which may be enough to produce the expected currency depreciation.

In the transactions table this change in preferences in the private sector is illustrated by the share for claims on the foreign sector ΔB_{FP} relative to claims on the banks ΔM_{BP}. This change in the portfolio distribution can be seen as further development of the Keynesian liquidity preference as it is influenced not only by the expected development of bond prices in the domestic market but also by the expected development of exchange rates.

The financial effect of the relations between cyclical fluctuations and the equilibrium of the balance of payments is of special interest as it shows clearly the relation between the production growth rate in the economy and the substitutive relations in the asset structure. As long as the growth rate is satisfactory there is a complementary relation between the growth of foreign assets and domestic liquid assets in the private non-bank sector and the bank sector. As soon as the stable growth rate is threatened and the

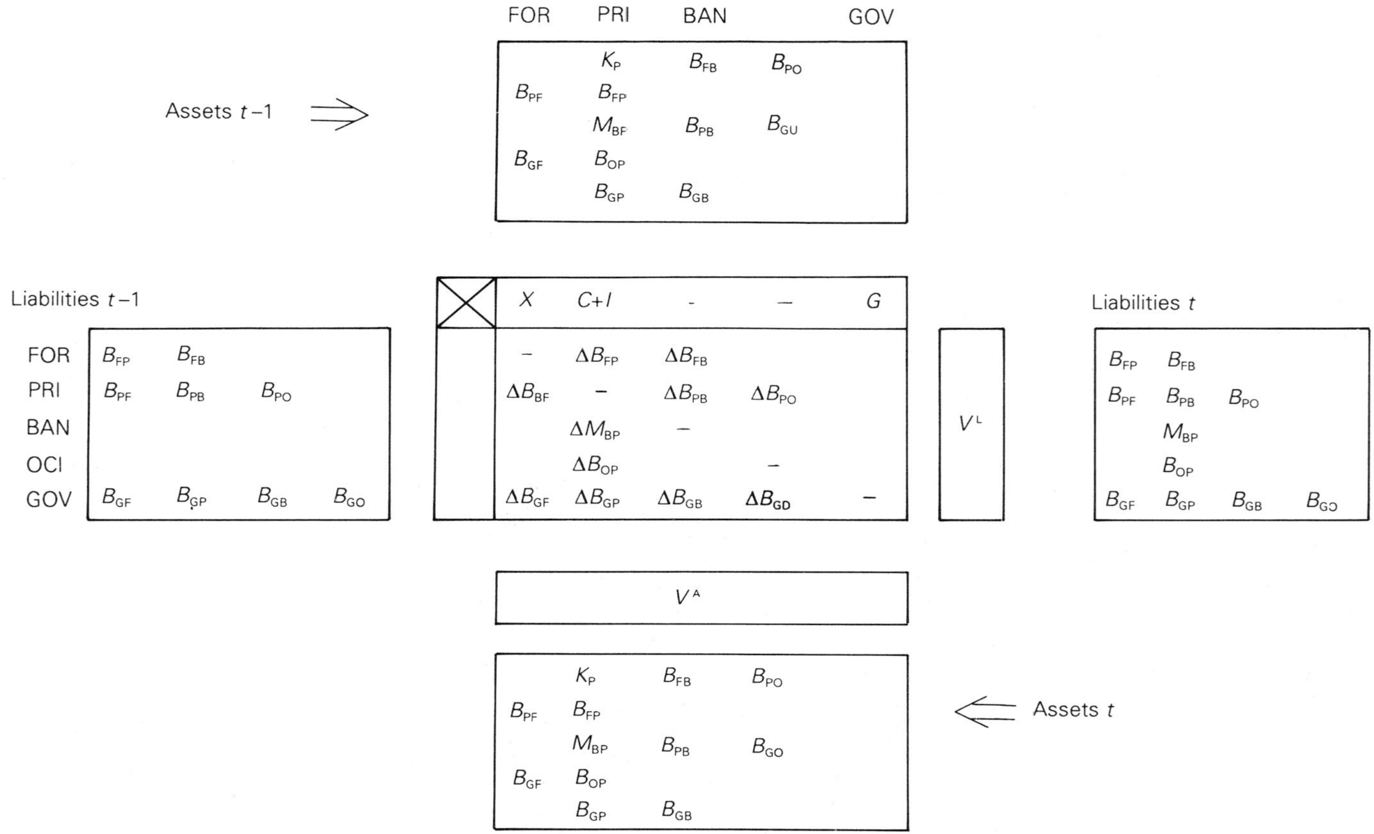

Fig. 7.1 Transactions table.

currency reserves do not develop according to expectations the complementary relations between assets are replaced by substitutive relations. The financial theory for this process was analysed in Chapter 6.

(c) *Stagnation and financial expansion: real growth and internal disequilibrium*

The high growth rate in major industrial countries during the 1950s and the 1960s was accompanied by financial expansion. This was the result of an expanding disequilibrium in real development. More dramatic was the increasing deficit in the balance of payments in the United States while the FRG and Japan built up accumulating surpluses. At the same time there was a continuous change in income distribution. As has been discussed in an earlier chapter, the profit share of value added in some industrial countries, including the FRG, Italy, the United Kingdom, Denmark, the Netherlands and Sweden, was decreasing from the middle of the 1950s to the end of the 1970s. Even though the deficit was increasing in non-financial enterprises in major industrial countries, the high growth rate in demand for investment in financial assets led to increasing equity prices. However, this development varied in different countries. In Japan industrial share prices increased continuously throughout the whole period from the beginning of the 1950s to the end of the 1970s, with a short interruption of falling share prices in the period 1973–4. In Sweden share prices were increasing. In the United States and the United Kingdom share prices peaked in 1972. In France and the FRG share prices were stagnant during most of the 1960s between a peak at the beginning of the decade and a second peak at the beginning of the 1970s.

When the growth rate sharply decreased in 1974 and became negative in most industrial countries in 1975 this development seemed to agree with Minsky's theories about the interrelations between real and financial development. His thesis, which we have cited in earlier chapters, is that a sustained growth process is usually accompanied by a financial weakening in both enterprises and households. In such a situation a normal cyclical recession may trigger off a financial crisis, and this may lead to a prolonged deflation with sinking capacity utilization. The cyclical fluctuation will show deeper recessions and the next peak may not reach the level of the previous peak. We shall now make a more thorough study of his theories in order to try to explain the long period of stagnation with low growth rate and high unemployment which was dominant in many industrial countries during the second half of the 1970s and the first part of the 1980s.

Minsky has analysed his theory in a number of papers. The relations between real and financial development are thoroughly treated in Minsky (1984, pp. 173–380). The theory starts from a medium-term growth process which is called a 'sustained boom' and is the result of increasing private investment or government demand. The sustained boom is interrupted by intermittent recessions due to supply constraints. These recessions are not equally severe. They vary in amplitude and duration, and in the interval of time until the previous peak income is exceeded. These differences in the severity of the recessions are due to differences in the financial environment in which they take place.

If a recession occurs when the private sector's portfolios include a high share of liquid assets and the decrease in the growth rate is expected to be short so that growth rate expectations for the medium-term development are unchanged, the financial condition will be a help for a revival of consumption expenditure and investment in inventory so that a revival of expenditure will very soon put an end to the recession. However, during a sustained boom the financial development is gradually changed. In an economy dominated by private expenditure four changes are of particular importance:

1. idle cash balances are decreased;
2. the amount of monetized private debt may increase;
3. the ratio of debt to equity rises;
4. debt payments rise relative to income.

In order to show how these changes will increase the chances that a decline in the growth rate will result in financial distress, which in turn will lead to prolonged recession, Minsky has employed a more formal analysis. He uses the Hicksian type of multiplier accelerator model which is written (Minsky, 1964, p. 202)

$$Y_t = C_t + I_t$$

$$C_t = \alpha_0 + \alpha_1 Y_{t-1} \qquad 0 < \alpha < 1$$

$$I_t = \beta_0 + \beta_1(Y_{t-1} - Y_{t-2})$$

Minsky also follows Hicks in his assumption that the marginal propensity to consume α and the accelerator β will have values such that normally the model will produce monotonically explosive time series. The two constants α_1 and β_0 depend upon the past and current states of the economy. They are used to feed the effects of financial changes into the model. There are two distinct phenomena which together transform the financial environment and through this the two constants α_1 and β_0:

1. how investment is financed;
2. how growth expectations will affect the value of equity and real assets.

When the national income increases so that Y_t is greater than Y_{t-1}, the increase in Y must be financed from the sources outside Y_{t-1}.

This way of looking at the financial relations associated with income changes is consistent with the flexible accelerator-multiplier approach toward income generation. (Minsky, 1964, p. 206)

To make the relations more precise we shall distinguish between planned income (*ex ante* income) and actual income (*ex post* income). The same distinction will be made for consumption, savings and investment. The consumption function gives consumption *ex ante*: $C_t = \alpha_0 + \alpha_1 Y_{t-1} = t_1$ (*ex ante*). It is assumed that consumption plans always are carried out: $C_t(ex\ post) = C_t(ex\ ante)$. No problems arise with the financing of the induced part of consumption $\alpha_1 Y_{t-1}$ as it is covered by the income in the previous period: $0 < \alpha_1 < 1$. However, the autonomous part may be greater than that part of the income in the previous period which is not used for induced consumption: $(1 - \alpha_1)Y_{t-1}(ex\ post)$.

We shall assume that α_0 depends on the financial resources of the households in the form of wealth, liquid assets and ability to borrow. We can write $S_t(ex\ ante)$ as the difference between $Y_{t-1}(ex\ post)$ and $C_t(ex\ post)$:

$$S_t(ex\ ante) = Y_{t-1}(ex\ post) - C_t(ex\ post)$$

Given the planned investment $I_t(ex\ ante)$,

$$Y_t(ex\ ante) = I_t(ex\ ante) - C_t(ex\ post)$$

For planned income in period t to be greater than the actual income in period $t-1$, it is necessary for planned I_t to be greater than planned saving: $I_t(ex\ ante) > S_t(ex\ ante)$, for $Y_t(ex\ ante) - Y_{t-1}(ex\ post) = I_t(ex\ ante) - S_t(ex\ ante)$. We find that planned investment exceeds planned savings with the value of the increase in national income. How can this part of the investment be financed?

We know that there are three resources for financing investments.

1. Internal finance: the accumulated savings of the business sectors are used. No claims are created. Networth and market value of equity rise as a result of the investment.
2. External direct finance: savings agents receive the issued liabilities in exchange for funds.
3. External indirect finance: credit institutes receive the issued liabilities in exchange for funds. If the credit institutes are banks these funds are money created by the banks. If the credit investments are not banks these funds have been lent to the institutes.

Therefore we find that, if the national income is to grow, liabilities in excess of planned saving must be issued and the buyer of these liabilities must *either* have surplus cash balances which are not needed for transactions and which are not the result of planned saving *or* have the ability to create money (which is the prerogative of the banks).

In order to draw down the cash balances investors consider the expected development of bond prices. If they expect that bond prices have already decreased by as much as they are likely to in the planning period, the cash holder will make his investment. If banks have liquidity and expect increasing incomes in the business sector they will create money by buying claims on investors.

It is possible to illustrate real and financial investments as a process in which non-financial enterprises, banks and households are involved. A simplified transactions table can be used (Fig. 7.2). The financing process illustrated in Fig. 7.2 starts with the banks agreeing to lend to non-financial enterprises in order to finance that part of planned investment I_t which is not financed by the NON sector's own saving. In this case the investment has been financed partly by drawing down surplus cash in the NON sector and partly by money creation in the BAN sector (ΔB_{NB}). An alternative way of financing would have been if the part ΔB_{NH} which was not financed by the willingness of the NON sector to draw down its liquid assets had been financed by the household. ΔM_{BN} represents the decrease in the NON sector's liquid assets as a result of financing part of the investment with accumulated savings from earlier periods.

The transactions table indicates that if the central bank does not increase the liquid assets of the banks, for example, by open market operations, the ability of the banks to continue the process of money creation will decline. The non-financial enterprises and households will also be limited as to how much they will be able to draw down their cash holdings. There will be difficulties in maintaining the financial expansion which is indicated by a reduced share of liquid assets and a high ratio of debt to equity.

However, as long as the sustained growth rate is stable these changes in the distribution of assets and liabilities may be hidden by the rising prices in equities. Minsky has emphasized the importance of the relations between equity prices and exported income growth:

> Assuming that the expectation that economic growth will take place becomes progressively stronger as a sustained boom continues, the

		NON	BAN	HOU
	X	I_t		C_t
NON	Y_N		B_{NB}	B_{NH}
BAN	Y_B	M_{BN}		
HOU	Y_H			

Fig. 7.2 Simplified transactions table (*ex ante*).

> value of those assets which are expected to participate in the income growth will increase relative to the assets which are not expected to participate. This means that equity and real estate prices will rise relative to debt prices, and this will occur in the absence of any rise in the price of income. (Minsky, 1964, p. 213)

He shows how the value of the equities rises when the expectations of growth in the return are spread through the economy. The present value of \$1 per year is $\$1/r_1$ where r is the rate of discount appropriate to perpetuities of this particular rise and uncertainty class. The equities can be treated as perpetuities because the enterprises have an expected life which is so long that for practical purposes it can be considered as infinite. If, instead of having a value of \$1 per year, the equity is expected to have a perpetual growth of $g\%$ per year and the expected return next year is \$1, the present value of the equity is $\$1/Cr - g)$. During the process of sustained growth the valuation of equities will grow from $\$1/r$ to $\$1/(r - g)$. Minsky also assumes that in addition to this increase in equity values a speculative boom may increase values still further. As soon as the expectations of further growth come to an end, the value of these assets will fall. The change in asset distribution from liquid assets to illiquid assets and from equities to debt will be apparent. This will create financial distress. When the need for liquidity forces the financial intermediaries to sell assets, financial distress spreads throughout the economy and produces a financial crisis.

Minsky has also analysed how a financial crisis influences expenditure on consumption and investment. He emphasizes the inertia of consumption and investment expenditure. Households will try to keep up consumption expenditure despite reduced incomes. This is possible as long as they can draw on their financial resources. These depend on the households' liquid assets and also on their ability to borrow. If the recession is prolonged, there is a risk that the financial intermediaries will have difficulties in continuing to give credits to households and consumption will have to decrease.

Investment expenditure may also be rigid and only adjust slowly to falling income and financial pressure. The type of investment which is most dependent upon income development is inventory investment. The decrease in inventories which takes place during the first part of the recession will lead to improved liquidity. In this way the need for adjustment can be postponed.

Financial pressure will be serious when lack of liquidity makes it difficult to meet financial commitments. In a financial crisis non-financial enterprises and credit institutes are in the same illiquid situation. Asset prices will fall partly as a result of pessimistic income expectations and partly because of increasing sales in order to improve liquidity. This discussion of some parts of Minsky's theory shows that there are some areas of financial development in industrial countries which are in accordance with his theory and help to explain the period of high sustained growth and the subsequent stagnation. These include the financial development during the period of sustained intermediaries such that there was an average increase in debt-to-equity ratios in industrial countries. However, this development can be explained by a decreasing profit share in some of the major industrial countries. Minsky does not discuss the importance of changes in factor income distribution.

The subsequent period of production stagnation was also accompanied by financial distress and some major international banks were in danger of insolvency. However, the financial pressure was more the result of restrictive monetary policy in industrial countries which had been made necessary by the large balance-of-payments deficits and the high inflation. This development is above all a consequence of foreign trade and international capital transactions between industrial countries—a problem which has not been addressed in Minsky's theory.

7.3 PUBLIC SECTOR AND ECONOMIC POLICY IN AN OPEN ECONOMY

In Chapter 1, Tables 1.1 and 1.2, we outlined a situation where the internal demand became less important for the domestic industry but its share of the international market was still low, implying a high market risk. This situation is normal whenever we have a production system which can utilize large-scale techniques and the domestic demand potential is limited. It is typical for small open economies. European countries in particular have faced this situation during the last two decades. Therefore it is important to understand the changing environment of economic policy.

In this section we shall analyse the conditions of economic policy in an economy which has two sectors. One has a high growth rate but is strongly dependent on production capital. However, this sector will face the short-run characteristics, with varying labour productivity and profit share according to variations in production, discussed in Chapter 2. The other sector has little or no possibility of increasing production by investing in new capital. Furthermore, we assume that the first high-productivity sector is exposed to international competition while the second sector is not.

According to the model proposed by Baumol (1970), which we referred to at the end of Chapter 2, the general predictions of such a model are that the costs of the sector with zero or low productivity growth will grow relative to the high-productivity sector.

There has been much argument about the increasing costs of public sector production and its decreasing productivity. A parliamentary investigation of the effectiveness and productivity of the Swedish public sector (SOU, 1988) found negative productivity growth in many public sectors. They define the cost productivity as

$$\text{productivity} = \frac{\text{output in physical terms}}{\text{total cost}} \tag{7.15}$$

They then discuss 'the mystery of negative productivity changes' (SOU, 1987, p. 57).

Assume a fairly effective labour market. It is then reasonable to believe that the average wage level will have a positive functional relationship with the productivity increase in the high-productivity sector, as Baumol assumes in his model, particularly if the low-productivity sector is financed through taxation. Using (7.15) we find

$$z^{\circ} = \frac{Q^{\circ}}{w(z^{P})L^{\circ}} \tag{7.16}$$

where z° is public sector productivity, z^{P} is private sector productivity and Q° is public sector production in physical terms. If then $Q^{\circ} = \zeta L^{\circ}$ we have the derivative of public sector cost productivity with respect to productivity in the private sector:

$$\frac{\delta z^{\circ}}{\delta z^{P}} = -\frac{\zeta w^{1}(z^{P})}{[w(z^{P})]^{2}} \tag{7.17}$$

If we transform the public sector into a competing market sector we shall of course have reduced production according to the Baumol model. This will then probably have effects on the welfare distribution.

The problem of the negative productivity change in public production should, in itself, be no 'mystery' to economists.[4] However, in an economy which is strongly dependent on foreign trade, this problem will create serious stability problems.

First, let us introduce the public sector into the dynamic model given by (2.25) in Chapter 2, still without a foreign sector:

$$\begin{aligned}
Q_1^{D} &= \psi_1\left(\frac{w}{P_1}L - wL_2\right) + \frac{\psi_2\pi}{P} \\
Q_1^{S} &= z(L - L_2) \\
L^{D} &= \phi \mathrm{MC} + L_2 \\
L^{S} &= \bar{L} \\
\dot{P} &= Y_1(Q_1^{D} - Q_1^{S}) \\
\dot{w} &= Y_2(L^{D} - L^{S})
\end{aligned} \tag{7.18}$$

[4] We repeat that the condition is that it is not possible to increase output per labour unit. We could even think of a negative change of output per labour unit while increasing quality. The piano concertos by Mozart are generally played at lower tempo nowadays compared with the time of Mozart, depending on the richer sound of modern grand pianos compared with the forte-pianos of Mozart's day.

The short-run production function has the same properties as we discussed in Chapter 2. We have a capital stock K^* which implies an optimal production volume Q^* at which MC attains its minimum. We must then discuss two different situations, i.e. when actual production is below or above Q^*.

We remind ourselves of the importance of the volume of capital stock. If the capital stock is small, the optimal production Q^* will occur at a low volume of employment. An expansion of demand to solve the employment problem will then quickly move the production into the interval $Q > Q^*$, which we proved in Chapter 2 to be an unstable interval if equilibrium is attained.

In the model we assume that the wages are determined by the cost and productivity conditions of the private sector 1. The demand Q_1^D for goods produced in the private sector is a function of the total real wage income (in both sectors) reduced by the wage costs of public production (we disregard depreciation of public capital stock) and the private investment demand determined by the real profit. The labour demand consists of both the private sector demand and the public sector demand. We look upon the latter as a policy variable. As in Chapter 2 we discuss two regimes: one when the actual production in sector 1 is below the optimum relative to the existing capital stock, and the other when it is above this value.

(i) *Regime 1:* $Q_1 < Q^*_1$. As in Chapter 2, Equation (2.30a), we insert the demand and supply functions into the excess demand functions. The expressions are rather complex and hence we make the simplifying definition

$$L_2 = \rho L \qquad P = \frac{L_2}{L} \qquad 0 < \rho < 1 \tag{7.19}$$

The control parameter of the authorities will then become ρ:

$$\begin{aligned} \dot{P} &= \frac{\gamma_1 L}{P}\,[w(\psi_1 - \psi_2) + P[\psi_2 z(1+\rho) - \psi_1 w\rho] + \psi_2 w\rho] \\ \dot{w} &= \frac{\gamma_2(\phi_1 z - L)}{w(1-s)(1-\rho)} \end{aligned} \tag{7.20}$$

It is now obvious that we do not *a priori* have a price–wage equilibrium at Q^*. If the complementary labour demand of the public sector adds up to the total supply of labour we shall have equilibrium; otherwise we shall not.

Solving the equilibrium $\dot{P} = 0$ and $\dot{w} = 0$, we have (disregarding trivial equilibria)

$$\begin{aligned} P &= \frac{w[\psi_2(1-\rho) - \psi_1]}{z\psi_2(1+\rho) - \psi_1 w\zeta} \\ w &= \frac{\phi_1 z}{\overline{L}(1-s)(1-\rho)} \end{aligned} \tag{7.21}$$

By comparing with the corresponding equilibria given by (2.34) and (2.35) we can draw the following interesting conclusions.

1. The introduction of a public sector will raise the equilibrium wage in the unemployment equilibria.
2. Then prices will no longer be given by a simple relation between wages and average productivity. The interesting thing is that when we introduce a public sector the relative elasticities of consumption and investment demand must be taken into consideration. If the elasticity of investment demand is low when there is unemployed capacity, which is probable, the price level at equilibrium will be lower the higher is the share of total employment.

Differentiating (7.20) with respect to prices and wages will give us the following stability conditions:

$$H'\rho(P, w) = -\gamma_1 \frac{L}{\rho^2}[\psi_2 z(1+\rho)] \tag{7.22a}$$

$$H'w(P, w) = \frac{y_1(\delta L/\delta w) - L}{p^2}[\psi_1(1-\rho) + \psi_2(\rho - 1)]\frac{\delta L}{\delta w} < 0 \tag{7.22b}$$

$$G_1'(\mathrm{w}) = -\frac{\gamma_2 \phi z}{w^3(1-s)^2(1-\rho)^2} \tag{7.22c}$$

The system is stable, which means that it is possible to increase demand by fiscal policy without raising the price. The price of the private sector will then be reduced when labour productivity increases.

Let us now expand the model in (7.22) by including foreign trade. To keep the model reasonably straightforward we must simplify the expressions and make them more abstract to minimize the numbers of parameters. The goods demand is given by

$$Q^{\mathrm{D}} = \frac{\theta}{p} f_{\mathrm{D}} + (1-\theta) f_1 + \psi_2 \frac{\pi}{p} \tag{7.23}$$

where

θ is the domestic share of total demand

$f_{\mathrm{D}} = f_{\mathrm{D}}(w, L, \rho)$ is domestic demand

p is the share of public employment in total employment

$f_{\mathrm{I}} = f_{\mathrm{I}}(p^{\mathrm{I}}/p, \sigma)$

P^{I} is the international price level

σ is a growth factor of international demand

ψ_2 is the elasticity of investment demand with respect to real profit

$\pi = P\theta f_{\mathrm{D}} + P_{\mathrm{I}}(1-\theta) f_{\mathrm{I}} - w\phi(\mathrm{MC})$

The marginal cost function depends on the inputs of imports:

$$\mathrm{MC} = \frac{w(1 - S)}{z} + \lambda \frac{P_{\mathrm{I}}}{P} \tag{7.24}$$

The goods supply is given by

$$Q^{\mathrm{S}} = Z(L - pL) = \frac{z\phi(\mathrm{MC})}{1 - \rho} \tag{7.25}$$

Using the same expressions as before for the labour market and excess demand function we have the following model:

$$\begin{aligned} Q_1^{\mathrm{D}} &= \theta f_{\mathrm{D}} + (1 - \theta) \frac{P_{\mathrm{I}}}{P} f_{\mathrm{I}} + \frac{\psi_2 \omega}{P} \phi(\mathrm{MC}) \\ Q_1^{\mathrm{S}} &= \frac{z\phi(\mathrm{MC})}{1 - \rho} \\ L^{\mathrm{D}} &= \frac{\phi(\mathrm{MC})}{1 - \rho} \\ L^{\mathrm{S}} &= \bar{L} \\ \dot{P} &= Y_1(Q^{\mathrm{D}} - Q^{\mathrm{S}}) \\ \dot{w} &= Y_2(L^{\mathrm{D}} - L^{\mathrm{S}}) \end{aligned} \tag{7.26}$$

We remember that when $Q < Q^*$ we have

$$\phi(\mathrm{MC}) = \phi_1 \frac{1}{\mathrm{MC}} = \frac{Pz}{Pw(1 - s) + \lambda P_{\mathrm{I}}} \tag{7.27}$$

and when $Q > Q^*$ we have

$$\phi(\mathrm{MC}) = \phi_2 \mathrm{MC} = \frac{Pw(1 - s) + \lambda P_{\mathrm{I}}}{Pz} \tag{7.28}$$

Even in an extremely simple model, like the one we use here, the introduction of international trade complicates the problem considerably, even though we have simplified the notation and reduced the number of parameters. The dynamic equations have the following shape if $Q < Q^*$, i.e. if there is unemployed capacity:

$$\begin{aligned} \dot{P} &= \gamma_1 [\theta f_{\mathrm{D}} + (1 - \theta) \frac{P_{\mathrm{I}}}{P} f_{\mathrm{I}} + \frac{\psi_2 \omega z}{w(1 - s) + \lambda P_{\mathrm{I}}} \\ &\quad - \frac{Pz^2 \phi_1}{Pw(1 - s) + \lambda P_{\mathrm{I}}(1 - \rho)} \\ \dot{w} &= \gamma_2 \frac{Pz\phi_1 - \bar{L}}{(1 - p)(P\mathrm{w})(1 - s) + \lambda P_{\mathrm{I}}} \end{aligned} \tag{7.29}$$

(ii) *Regime 2:* $Q > Q^*$. If $Q > Q^*$, i.e. if the actual demand is greater than expected, we have the corresponding functions

$$\dot{P} = \gamma_1[\theta f_D + (1 - \theta) \frac{P_I}{P} f_I + \frac{\psi_2[w(1 - s) + \lambda P_F]}{wz} - \frac{\phi_2[Pw(1 - s) + \lambda P_I]}{P(1 - \rho)}$$

$$\dot{w} = \frac{\gamma_2[(Pw(1 - s) + \lambda P_I)\phi_2 \bar{L}]}{Pz(1 - \rho)} \tag{7.30}$$

Expressions (7.29) and (7.30) are rather complicated and we have to make some simplifications. If we first look at the wage functions we find the following conditions:
for $Q < Q^*$

$$w = \frac{z\phi_1}{\bar{L}(1 - \rho)(1 - s)} - \frac{\lambda P_I}{P(1 - s)} \tag{7.31}$$

and for $Q > Q^*$

$$w = \frac{\bar{L}z(1 - \rho)}{\phi_2(1 - s)} - \frac{\lambda P_I}{P(1 - s)} \tag{7.32}$$

The more important imported goods are as inputs in domestic production the more dependent on foreign prices the equilibrium wages will be. This dependence will be negative in both cases, but there is a difference between the two situations. When $Q < Q^*$ the elasticity of labour productivity with respect to production is positive. This implies that the negative effect of foreign prices on wages will be more important the higher the value of the elasticity. However, we shall have the opposite effect when $Q > Q^*$ because of the negative value of s.

We can also see from the two equations that the share of public employment P will have a positive effect on the equilibrium wage when the capacity utilization is low and a negative effect when production is above the optimum level. To find the properties of the local equilibria we set $P = H(p, w)$ and $w = G(p, w)$. To simplify notation we just indicate the signs of the derivatives. In matrix notation the system is as follows:

$$\begin{vmatrix} \dot{P} \\ \dot{w} \end{vmatrix} = \begin{vmatrix} H'p & H'w \\ G'p & G'w \end{vmatrix} \begin{vmatrix} P - \hat{P} \\ w - \hat{w} \end{vmatrix} \tag{7.33}$$

where $\hat{P}$ and $\hat{w}$ are the equilibrium values. The local stability is determined by the signs of the matrix of the derivatives. If both roots have negative real parts the system will be locally stable. We obtain the signs from the equation

$$\mu^2 + (H'p + G'w)\mu\Delta = 0 \tag{7.34}$$

where Δ is the determinant. If the determinant is positive the local stability of the system will be determined by the sign of $(H'p + G'w)$. If this is negative the system is locally stable; if it is positive, the system is unstable.

From (7.30) we find

$$\begin{aligned} H'p &< 0 \\ H'w &> 0 \\ G'p &> 0 \\ G'w &< 0 \end{aligned} \tag{7.35}$$

since $f'_{\mathrm{D}}p < 0$. When there is unemployed capacity $\psi_2 \approx 0$ and $s > 0$ and we can make the approximation $H'w \approx \gamma_1 \theta f_{\mathrm{D}} w$. If $|H'pG'w| > |H'w\,G'p|$, which is most probable here, there will be one positive and one negative sign which implies a saddle-point equilibrium. Unlike the closed economy there will be no stable equilibria when capacity is underutilized.

From system (7.31) we obtain

$$\begin{aligned} H'p &< 0 \\ H'w &< 0 \\ G'p &> 0 \\ G'w &> 0 \end{aligned} \tag{7.36}$$

The system will have no stable equilibria if $G'w > H'w$; otherwise it will have saddle-point equilibria. Thus we find that the introduction of foreign trade completely changes the stability properties of the model compared with those of the closed economy. These analytical results correspond well to the experiences of stabilization policy in many countries which have become increasingly more dependent on the international development.

REFERENCES

Baumol, W. J. (1970) *Economic Dynamics*, Macmillan, New York.

Eichengreen, B. and Parkers, R. (1987) The anatomy of financial crises. Seminar Paper 375, Institute for International Economic Studies, University of Stockholm.

Kragh, B. (1967) *Finansieringsproblem och Strukturomvandling*, Prisma, Stockholm.

Minsky, H. P. (1964) Financial crises. financial systems and the performance of the economy, in *Private Capital Markets* (eds I. Friend, H. Minsky and V. Andrews), Prentice-Hall, Englewood Cliffs, NJ.

Sachs, J. (1985) External debt and macroeconomic performance in Latin America and East Asia. *Brookings Papers on Economic Activity*, **2**

SOU (Statens Offentliga Utredningar) (1987) 3 Långtidsutredningen Bil. 21 Den offentliga sektorn–produktivitet och effektivitet. Allmänna Förlaget, Stockholm.

(Parliamentary Report 1987: 3 App. 21 The Public Sector – Efficiency and Productivity.)

Surrey, M. (1982) United Kingdom, in *The European Economy: Growth and Crises* (ed. A. Boltho), Oxford University Press, p. 537.

Tinbergen, J. (1966) *Economic Policy: Principles and Design*, North-Holland, Amsterdam.

van der Hoeven, W. (1987) Strukturförändringar inom bankverksamheten inom OECD. *Kredit- och Valutaöversikt*, **4**.

Haaland, J. I., Norman, V. D., Tutherford, T. and Wergeland, T. (1986) *VEMOD –a Riccardo-Heckscher-Ohlin-Jones Model of World Trade*, Working Paper MU 11/86. Centre for Applied Research, Norwegian School of Economics and Business Administration, Bergen.

8

Concluding remarks

The main theme of this book has been economic disequilibrium. The fundamental meaning of this concept in a dynamic context is that the agents are unable to realize their plans. Therefore we have found it suitable to discuss disequilibrium using the *ex ante–ex post* concepts of the Stockholm School.

'Disequilibrium economists' like Lejonhufvud, Malinvaud and Grandmont, to mention some influential followers of Keynes, have recognized economic disequilibrium as a qualitatively different situation from equilibrium in that the agents will behave according to the relevant decision variables in a way which is non-optimal from an equilibrium point of view. Disequilibrium is generally analysed as an equilibrium where the agents are rationed to some extent. In a mathematical model of the economic behaviour of agents this would imply parameter changes in the model. In general, however, it is assumed that equilibrium will be restored when the imposed restrictions are removed; the disequilibrium is reversible.

In general equilibrium, in addition to the uniqueness of the price vector, there is also a unique relation between the price vector and the distribution of endowments at the time of exchange. This implies that, when trade occurs under rationing, the distribution of endowments will change. We could then talk about 'false' trading relative to the equilibrium price vector. Given the changed distribution of endowments there exists an equilibrium, but this equilibrium will be different from the one existing before the rationing occurred provided that the utility functions of the agents are not of an identical character. Disequilibrium must then create irreversible redistributions, in the sense that there are no market processes fulfilling the neoclassical conditions which will make the economy to return to its original equilibrium.

Economists often characterize Keynesian theory as a short-run disequilibrium theory while they maintain that neoclassical theory describes the long-run development of the economy. We interpret this to mean that in the short run there will exist the possibility of disequilibria or temporary equilibria under rationing, but in the long run the economy will converge to some unique equilibrium path. Some economists even claim that the basic condition for this is that the markets are relieved from the interference of collective actions. We maintain that this is a logical fallacy. If one accepts the possibility of short-run disequilibria, one should also accept that the trading-out of equilibrium will change the conditions for future market processes.[1]

Our approach is that neoclassical theory is generally valid as a description of *ex ante* behaviour. The general equilibrium theory provides the conditions when, *a priori*, *ex ante* will be equal to *ex post*. If the conditions are not fulfilled, a new distribution of endowments will arise from the market. This will in turn become the conditions for further *ex ante* actions.

The conditions for general equilibrium in production economies are demanding with respect to information and the flexibility of productive resources. In an empirical world they would not normally be fulfilled. However, during periods of social and political stability and generally increasing demand, equilibrium paths could exist locally. These are broken when the demand changes its direction or the technological conditions change. Deficient flexibility, of production capital in particular, will promote the kind of changes that will throw the economy out of equilibrium through revaluation of the existing capital stock.

On p. 13, Fig. 1.5, we introduced an extended input–output table. In addition to the real transactions it also includes the financial transactions necessary to transform factor incomes into demand for goods and services. This transactions table can be seen as an instantaneous temporary equilibrium on the real and financial markets. The instantaneous pictures are connected to each other in the time dimension through the stock of assets and liabilities, Fig. 1.11, p. 26. To complete this connection we have to allow for revaluations of the existing stocks of assets which is done through the V vector.

If the economy develops along some equilibrium growth path it will imply that the coefficients of the financial matrix F are constant and the revaluation vector V is zero. The stocks will then increase through continuous additions from the factor incomes and basically the financial stocks will be in a complementary relation with physical capital. The occurrence of revaluations indicates disequilibria in the real sector.

In our analysis we stress the importance of factor income distribution. If it is changed, the aggregate behaviour of the financial market will change and we shall generally have revaluations of assets.

It is important to realize that when the economy is on an equilibrium path the financial matrix will be of little interest since the coefficients are

[1] Another interpretation of long-run equilibrium is that the competition will result in development of the economy into a situation where all economic resources are equally distributed and the possibility of growth vanishes. The concept of economic equilibrium will then be similar to that of thermodynamic equilibrium. (Lavois, 1983)

stable. This implies that the creditors' and debtors' expectations of the future converge.[2]

The fundamental role of the financial markets is to transform savings into productive investments. However, this role is perverted by the possibility of storing value for the future in non-productive money and financial contracts. This focuses interest on the role of money. Sometimes money is assigned an almost metaphysical role in discussions of economic disequilibria. The role of money as a store of value is due to the fact that it is socially accepted as a medium of exchange. This makes money independent of the distribution of preferences over the goods space. Let us return to the Arrow–Debreu world. A consumer is defined as a pair ($\lesssim w$), where $\lesssim$ is a strictly convex, continuous and complete preference ordering and w is a vector comprising partly initial endowments and partly shares of the production result (Debreu, 1959). Debreu (1974) shows that the elements of the endowment vector of an asset have to be present in at least one other agent's preference space in order to have positive prices. This implies that in a non-monetary world the agents must choose, as stores of value, goods which are present in as many different preference spaces as possible in order to have good security for the future. Here we have the traditional investment problem.

As long as money is socially acceptable as a medium of exchange, it will minimize the uncertainty of the future value of wealth. Money is then independent of the future preference structure. In periods when uncertainty about the future preference structure is important agents will choose the medium of exchange as a store of value, but this choice is unproductive in two ways. First it reduces the current demand for goods (which then affects the present factor incomes and rates of return on capital), and second it reduces the future supply of goods.

The value of financial assets generally depends on how close they are to money. When they are close to goods such as financial claims on some specific production organization they will be closer to real assets in times of uncertainty. When they are claims on government they will become closer substitutes to money. If an agent is sure that a particular good will be demanded in the future, of course he chooses investment goods attached to the production of that particular good as a store of value. The role of money as a store of value will be diminished.

The problem of money is not unique to the monetary economy. In a

[2] Arrow and Hahn (1971, Chapter 14) discuss a small model of utility and profit maximization over two periods, when prices in period 2 are uncertain when decisions about investments and lendings are made. The lending of the households in period 1 is their fortune in period 2. The result of the model is as follows: 'If the creditor has a less sanguine expectation of future profit than did the debtor, there will be a discontinuous fall in the creditor's wealth'. Furthermore they conclude: '... when the debtor is bankrupt, equilibrium in the money market implies an excess demand for goods, while if he is not bankrupt, it implies an excess supply of goods, where there are "full employment" excess demands'.

barter economy the search for suitable stores of wealth is of equal importance. Liquidity in a barter economy must mean that the individual can always at least meet the basic needs of survival.

To understand the value of money it must be attached to the question of lexicographic preferences. This specific kind of preference structure is basic to the demand for liquidity, whether it is a monetary economy or a barter economy. It is the existence of preferences of this type that makes agents choose alternatives which are inferior in a long-term context.

In the capital asset pricing model proposed by Markowitz, Fama and Merton among others, the opposite view is taken when financial markets are discussed. By using the dynamic programming technique they disregard the instantaneous liquidity restrictions and let the long-term targets determine the present choice between consumption and investment. This type of financial analysis will never capture the essentials of the Keynesian problem of liquidity preference.

The neoclassical assumptions regarding the rational behaviour of the agent probably give a good description of *ex ante* behaviour, except for the disregard of instantaneous liquidity restrictions created by the commitments inherent in the stock of assets and liabilities. When uncertainty about the future is combined with lack of flexibility of productive resources, the *ex ante* plans which are carried through will turn into an unexpected *ex post* situation where rational behaviour has to be revised. Uncertainty about the future is basic to all human activities.

It is important that we do not confuse the concept of uncertainty with the concept of risk. With respect to certainty the risk situation is less certain, but in the real world the risk situation, where outcomes and probability distributions are known at least subjectively, is normally the best deal we can obtain. The normal situation is uncertainty about the future in the sense that the set of possible outcomes is not known.

In our study of economic disequilibria we have focused our attention on the following problems. Lack of flexibility in short-run production creates on the one hand changes in the volume of factor incomes and on the other hand changes in the distribution of factor income. The latter effect in particular will create changes in the behaviour of the financial markets, thereby affecting future liquidity restrictions. The changes in the volume and distribution of the factor incomes also affect both the valuations of the stock of production capital and the volume of investments.

When analysing production we stressed the fact that short-run production functions could be derived from long-run functions only under special circumstances. In the short run the relationship between labour and production implies a lower elasticity of labour demand with respect to changes in production. This implies the existence of short-run locally stable unemployment disequilibria at a lower real wage than in the optimum utilization of the capital stock. We showed that when the technology becomes more capital intensive, a higher risk of capital is implied because there will be greater sensitivity to variations in demand. This increase in capital risk will decrease the growth in the capital stock at the same time as the return on capital increases.

If the economy is closed the introduction of a public sector will stabilize the situation, but if the economy is strongly dependent on foreign trade the introduction of a public sector will increase domestic prices and wages and create a deteriorating competitive strength on the foreign market if its productivity has a low increase. When the country is strongly dependent on foreign trade there will be a conflict between competitive strength and social welfare provided by public services if the economy is to maintain full employment.

REFERENCES

Arrow, K. J. and Hahn, F. (1971) *General Competitive Analysis*, Holden-Day, San Francisco, CA.

Debreu, G. (1959) *Theory of Value*, Wiley, New York.

Debreu, G. (1974) Excess demand functions. *Journal of Mathematical Economics*, **1**.

Lavois, M. (1983) *Economie Appliquée*, **36**, 2–3.

Index of names

Subject index